Lei Chen
Jiahuang Ji
Zihong Zhang

Wireless Network Security

Theories and Applications

Lei Chen
Jiahuang Ji
Zihong Zhang

Wireless Network Security

Theories and Applications

With 62 figures

Editors
Lei Chen
Department of Computer Science
Sam Houston State University
1803 Avenue I, AB1-214
Huntsville, TX 77341, USA
E-mail: chen@shsu.edu

Jiahuang Ji
Department of Computer Science
Sam Houston State University
1803 Avenue I, AB1-214
Huntsville, TX 77341, USA
E-mail: jiahuang@shsu.edu

Zihong Zhang
NASA Johnson Space Center
2101 NASA Road One
Building 16, Rm 2008
Houston, Texas 77058
E-mail: zihongzh@gmail.com

ISBN 978-7-04-036680-8
Higher Education Press, Beijing

ISBN 978-3-642-36510-2 e-ISBN 978-3-642- 36511-9
Springer Heidelberg New York Dordrecht London

Library of Congress Control Number: 2013931236

Printed on acid-free paper

Springer is part of Springer Science+Business Media (www.springer.com)

Preface

This edited book, *Wireless Network Security: Theories and Applications*, aims to present a full picture of the state of the art research findings and results in the area of wireless network security. University professors, graduate students, researchers, and professionals in the areas of Wireless Networks, Network Security and Information Security, Information Privacy and Assurance, and Digital Forensics may find this book valuable and beneficial.

Our idea of writing and collecting chapters for a book on Wireless Network Security emerged in 2009. While books and articles on wireless networks and those on network security could be easily found in the market, online libraries, and databases, we could hardly identify a high quality collection of contemporary research on the security of different types of wireless networks. Researchers in this field found it difficult to correlate and compare the various security concerns and solutions in similar yet different wireless networks, such as Wireless Mesh Networks and Wireless Sensor Networks. We consider that setting a forum of discussions about the security of eight different types of wireless networks by the top researchers from both academia and industry in the U.S. and China would be an enjoyable gift to the literature.

Another motivation of this work was to prepare an excellent textbook for the proposed graduate course Wireless Network Security at Sam Houston State University (SHSU) in Texas, USA, where two of the three editors of this book had full time job as computer science professors. We hope that the current master students and future doctoral students in the department of computer science at SHSU will find this book easy to read and digest the knowledge within it. We also have the faith that this book will be accepted and favored, as textbook or reference, by universities, research institutions, and companies in China, USA, and many other countries in the world.

Acknowledgements

We would like to give thanks to all the members who served in the Editorial Advisory Board (EAB) for providing helpful thoughts and suggestions on the content and capacity of the book, as well as for reaching out and inviting potential authors of chapters and reviewers for peer evaluations. The EAB consists of (in alphabetical order by surname):

- Dr. Jianer Chen, Texas A&M University, TX, USA
- Dr. Peter Cooper, Sam Houston State University, TX, USA
- Dr. Wen-Chen Hu, University of North Dakota, ND, USA
- Dr. Chung-Wei Lee, University of Illinois at Springfield, IL, USA

- Dr. Jie Wu, Temple University, PA, USA
- Dr. Zhijun Wu, Iowa State University, IA, USA
- Prof. Lianhua Xiao, Natural Science Foundation of China (NSFC), China
- Dr. Qing Yang, University of Rhode Island, RI, USA
- Dr. Wei Zhao, University of Macau, Macau, China

All three editors would express our appreciation to Mr. Christopher Hale and Mr. Shane Ulbricht who helped proofread some of the chapters in the book.

Lei would like to give his earnest thanks to his beloved wife Bo Dai and adorable son Ziyue Ethan Chen for their understanding and support as their husband and father had to focus on this book in the countless nights and weekends.

Jiahuang would like to express her appreciation to her husband, Jiahua Yang, for his understanding during the time when she should have been with him but instead worked on the book.

Zihong would like to send his thanks to his lovely wife Susan Song for her supportive efforts.

Lei Chen, Jiahuang Ji, and Zihong Zhang
November 2012

Contents

Chapter 1
Applications, Technologies, and Standards in Secure Wireless Networks and Communications

Lei Chen[1]

Abstract

Wireless networks and communications are becoming an integrated part of people's everyday work and life. Mainly due to the nature of unconfined signal transmitted in air and relatively limited computation power and battery resources, designing and implementing security mechanisms in wireless networks is much more challenging than in wired environments. This chapter provides an overview of the applications, technologies, and standards of secure wireless networks and communications. As each type of wireless networks has its own characteristics and supports different applications, the technical details of security goals, technologies, and standards are elaborated in each of the following chapters.

1.1 Introduction

Since the year of 1880 when Alexander Graham Bell and Charles Sumner Tainter invented and patented the Photophone, human beings have not stopped discovering new and efficient means of wireless transmissions and communications[1]. In the year of 1898, Marconi encoded alphanumeric characters using analog signals and sent them as wireless telegraphs across the Atlantic Ocean. Not until 1957 was the first man-made satellite launched by the Soviet Union which was mainly for military purposes. In the following five decades, thanks to military and commercial demands, new wireless technologies and applications, such as radio, television, mobile phones, Global Positioning System (GPS), Worldwide Interoperability for Microwave Access

1 Sam Houston State University, Huntsville, TX 77341, USA. Email: chen@shsu.edu.

(WiMAX), Bluetooth, Wi-Fi, and sensor networks, emerged like mushrooms springing up after rain.

Regardless of the various technologies for wireless networks and communications, the trends of these networks and communications require higher bandwidth (data rates), enhanced security and reliability, as well as better convenience and reduced costs. For example, in cellular networks, Third Generation (3G) networks support data rates at hundreds of kilobits per second (kbit/s) while Fourth Generation (4G) increases them by ten times over larger geographical areas when using WiMAX (strictly speaking WiMAX is considered to stretch across both 3G and 4G)[2]. People not only hope to have always-on reliable Internet access at both work and home, but also anywhere, even at locations where no existing infrastructural network is available.

While it is common practice to apply cryptographic encryptions and integrity checks to data and management messages that can help secure the communications over wireless networks, it is extremely difficult to implement uniformed security due to a number of reasons. Wireless mobile devices may have extremely different computation power, battery power, input and output methods supported, frequencies used and media access controls, and signal encoding and modulation technologies. Therefore, it is not possible to have unified security standards across all wireless networks. For example, a Bluetooth headset without alphanumeric input or display can only support very limited and primitive authentication methods whereas a powerful wireless mobile workstation supports almost all security mechanisms found on desktop computers.

This book addresses the security goals of wireless networks and communications, discusses the relevant security technologies developed, and the corresponding security standards proposed and set forth for Wireless Cellular Networks, Wireless Local Area Networks (WLANs), Wireless Metropolitan Area Networks (WMANs), Bluetooth Networks and Communications, Vehicular Ad Hoc Networks (VANETs), Wireless Sensor Networks (WSNs), Wireless Mesh Networks (WMNs), and Radio Frequency Identification (RFID). The security in each of these types of wireless networks and communications is discussed in a designated chapter from 2 to 10, with Security in WSNs over two continuous chapters.

The purpose of Chapter 1 is not to present rich technical details of various mechanisms, technologies and standards for securing wireless communications, rather it gives an overview and summary of what are currently available to achieve the security goals and direct audience to the relevant chapter(s). In each of the following sections of this chapter we overview the security of a different type of wireless networks in terms of applications that require security, and technologies and protocols designed for securing these networks.

1.2 Overview of Security in Cellular Networks and Communications

As smartphones become prevailing in the cellular market the ever-growing demands of user data volume and bandwidth can never be satisfied. With the support from 3G and 4G cellular networks, a smartphone can not only make phone calls, but also complete tasks that used to be only available on desktop computers. With a smartphone, a user can receive, read, and send emails, browse the Internet, have audio/video conferencing, manage bank and stock accounts, etc. Many such daily applications require cellular networks and devices to provide data confidentiality, integrity, and authentication, e.g. checking balance of bank account or making a transfer require mutual authentication between the user (and/or device) and the server. All user data must be secured against unauthorized viewing or tampering.

There were no technologies for securing voice in the First Generation (1G) of cellular networks. Authentication and data encryption services were introduced in the Second Generation (2G) against wire-tapping. For such purpose, a Subscriber Identity Module (SIM) card was designed and placed in a mobile phone to hold the shared secret key which is used for one-way authentication from the Base Station (BS) to Mobile Device. Mutual authentication became required in the 3G along with improved security algorithms for both data confidentiality and integrity.

The algorithms used for authentication in cellular networks are A3 and A8, both of which are based on cryptographic algorithm COMP128. Although COMP128 was found broken, two later versions COMP128-2 and COPM128-3 have not been subject to cryptanalysis. The data encryption algorithm A5/3 overcomes the weakness of previous versions and is considered a strong encryption algorithm for Global System for Mobile Communications (GSM). In 3G networks, data confidentiality is protected using algorithm f8 and integrity is preserved using algorithm f9, along with nine other algorithms most of which are based on KASUMI cipher. For more technical details of these ciphers and security standards used in cellular networks, please refer to Chapter 2 of this book.

1.3 Overview of Security in WLANs

Compared to cellular networks, WLANs provide shorter communication range, e.g. 100 m (328 feet). Since a WLAN can be considered as simple as a LAN attached with a wireless Access Point (AP), it indicates that the security expectation of WLANs is same as that of LANs. In other words, WLANs need to support all security applications in daily work, life, and entertainment, such as downloading large volumes of confidential data, streaming High Definition (HD) video and audio in a confidential CEO meeting. The security requirements of WLANs essentially fall into the following five areas: confidentiality, authentication, access control, integrity, and intrusion detection and prevention.

A number of IEEE 802.11 standards specify security requirements of WLAN. Wired Equivalent Privacy (WEP) utilizes RC4 encryption algorithm, Cyclic Redundancy Code-32 (CRC-32) checksum algorithm, and a pre-established shared secret key (base key) to encrypt the transmission between the clients and APs. Mainly due to short 24-bit Initial Vectors (IVs) used in WEP, it suffers a number of attacks and therefore is considered deprecated, although still available in most of the wireless routers found in the current market. Wi-Fi Protected Access (WPA) not only supports Pre-Shared Key (PSK) but also provides, in its enterprise mode, industrial level security using Remote Authentication Dial In User Service (RADIUS). The Temporal Key Integrity Protocol (TKIP) in WPA helps overcome the vulnerability found in WEP. Chapter 3 of this book provides more details of the technologies and standards in WLANs.

1.4 Overview of Security in WMANs

Wireless Metropolitan Area Networks (WMANs) provide wireless communications at acceptable bandwidth over much larger geographical areas compared to WLANs. WMANs use WiMAX technologies to provide Mobile Stations (MS) communications with Base Stations connected to backbone networks and the Internet. Also known as the "last mile" technology, WiMAX was designed and developed to have relatively long communication range, e.g. 8 km (5 miles), that fits wonderfully in urban areas. WMANs support all secure applications that can run over the Internet.

Also known as Wireless Local Loop (WLL), WMANs are based on the IEEE 802.16 standards with commercial name WiMAX. With its global market growing rapidly in recent years, WiMAX is becoming a major competitor among the prevailing wireless communication technologies. For secret key and data confidentiality, WiMAX uses RSA, Data Encryption Standard (DES) in Cipher Block Chaining (CBC) mode, or DES-CBC, and AES in Counter with CBC-MAC (CCM) mode. Data integrity is implemented using Hashed Message Authentication Code (HMAC) and Cipher-based Message Authentication Code (CMAC). Entity authentication is based on digital certificates and Secure Socket Layer/Transport Layer Security (SSL/TLS), and message authentication is done by using HMAC. Although improved IEEE 802.16 standards and amendments were published and adopted in almost every year of the past decade, existing standards still contain a number of security vulnerabilities inherent from deprecated versions. Chapter 4 of this book explains in details how these security goals are achieved in WMANs and what security vulnerabilities, threats and countermeasures exist.

1.5 Overview of Security in Bluetooth Networks and Communications

In contrast to most other types of wireless networks, Bluetooth communications are commonly found between two or more closely located, e.g. within 10 m (33 feet), Bluetooth enabled devices. Bluetooth applications may require secure authentication, data encryption and integrity check. For example, with multiple Bluetooth capable devices in the same public areas, secure authentication and voice data encryption should be enforced between a smartphone and the paired headset. However, compared to other types of wireless networks, applications over Bluetooth are relatively simple and data volume is much less and therefore the security complexity is also reduced.

Bluetooth is not based on the Internet Protocol (IP) and therefore cannot make use of the advanced IP-based standard security features and standards, such as SSL/TLS, digital certificates, or IP Security (IPSec). In order to meet various security requirements, four security modes were designed and implemented, and each Bluetooth device must operate in one of these four security modes. Each Bluetooth version supports one or multiple (not all) security modes. Bluetooth Authentication makes use of a challenge-response scheme for the verifier to identify the claimant. Successful authentication indicates that the claimant possesses the shared Link Key which is used for encrypting communication data thereafter. Data confidentiality is implemented by Exclusive-ORing plaintext with keystream, and therefore the generation of keystream is of importance. For more details of how keystreams are generated and other Bluetooth security technologies and standards, please refer to Chapter 5 of this book.

1.6 Overview of Security in VANETs

While there are a number of characteristics that distinguish VANETs from other types of Mobile Ad Hoc Networks (MANETs), two of them are considered most relevant: a VANET supports both Intervehicle (V2V) and Vehicle-to-Roadside (V2R) communications, and high mobility of vehicles with constraints of road topology. All applications that require security over the Internet are expected to be supported by VANETs. Thanks to security considerations and implementations in VANETs, users are able to run applications with confidential data in a vehicle moving at 120 km/h (75 mph) just like at home or at work.

A Public Key Infrastructure (PKI) with Certification Authority (CA, a Trusted Third Party, or TTP) is used to introduce trust within the network. The trial-use standard IEEE 1609.2 (previously named P1556) also addresses security services for VANETs. This standard targets the issues of securing Wireless Access for the Vehicular Environment (WAVE) messages against eavesdropping, spoofing, and other attacks. Based on industry standards for public key cryptography, the components of the IEEE 1609.2 security infrastructure include support for Elliptic Curve Cryptography (ECC), WAVE certificate formats, and hybrid encryption method. For more details of how to secure VANETs using these security components, refer to Chapter 6 of this book.

1.7 Overview of Security in WSNs

Wireless Sensor Networks (WSNs) are a type of MANETs that consist of a large number of resource-constrained sensor nodes. The flexibility in deployment and maintenance advances WSNs' applications in many fields, including military, environmental monitoring, public safety monitoring, emergency handling, medical and oceanic monitoring. For example, WSNs can be used to detect and track the intrusion of enemies or their tanks in a battle field, to detect forest-fires and floods, to monitor environmental pollutions, or to measure traffic flows in a traffic network. Security is one of the most important issues in WSNs mainly because WSNs are usually deployed in hostile or remote environments and work in an unattended manner.

Many schemes and algorithms were proposed in the recent research in different aspects of WSNs security, such as key management, secure routing, location privacy, secure data aggregation, attack defense, trust management, etc. Although there is no existing standard specifically developed for WSNs, several standards specify the technical requirements for other types of networks, e.g. wireless personal area networks, that can be considered as applications or variations of WSNs. Standard cryptographic ciphers, such as Digital Signature Algorithm (DSA) and ECC, can also be applied to secure WSNs. This book includes two chapters (7 and 8) to address the security issues and solutions in WSNs.

1.8 Overview of Security in WMNs

A typical Wireless Mesh Network (WMN) is a collection of WLANs that are interconnected to form a meshed WLAN. For this reason, WMNs are expected to support same applications and provide same level of security as WLANs. Application data originally run through WLANs then through LANs which are connected to the backbone networks can now be routed via other connected WLANs. When smartphones are supported by WMNs, voice conversations, web browsing, data transactions, or cloud clients' extensive access to cloud services will now be directed to more reliable and manageable mesh of WLANs.

Besides the security issues in WLANs, WMNs also have other security concerns such as secure routing and utilize additional protocols like the Simultaneous Authentication of Equals (SAE). To date the most important standardization process for Wi-Fi-based WMNs is the IEEE 802.11s. However, at the time of this writing, it is still in the draft development phase. Open 802.11s is a project to closely monitor the standardization progress of IEEE 802.11s and implement its functions faithfully in the open source Linux operating system. Chapter 9 of this book provides more details about the security technologies and future standards in WMNs.

1.9 Overview of Security in RFID Networks and Communications

Radio Frequency Identification (RFID) plays an important role in the proposed Internet-of-Things (IoT), an Internet-like structure that virtually presents all uniquely identifiable objects (things)[3]. RFID technology consists of small, inexpensive, computational devices, with wireless communication capabilities. Currently, the main application of RFID technology is in inventory control and supply chain management fields where RFID tags are used to tag and track physical goods. Security is important in RFID applications. For example, a thief armed with an RFID reader can wirelessly scan belongings of people close by and target wealthy ones with expensive items.

The standards for RFID-enabled passports are maintained by the International Civil Aviation Organization (ICAO), which maintains, among other things, the protocols needed to access the RFID tag embedded within passports. In recent years, there have been numerous RFID security protocols proposed and new RFID vulnerabilities discovered. The difficulty in securing RFID lies in the resource constraints of the RFID tags, which makes it impossible to adopt existing security solutions from other fields such as mobile computing or wireless networking, onto RFID networks. Chapter 10 of this book studies the security of RFID networks by first discussing background on RFID networks, followed by an introduction to main RFID threats. The chapter also reviews and analyzes basic RFID security protocols, then discusses more on advance attacks and defense, as well as the security of industry standard RFID protocols.

1.10 Summary

Although each different type of wireless network has its own characteristics, applications with security concerns typically require (mutual) authentication, data (and control message) confidentiality, and data (and control message) integrity. In addition to these security goals, security related information, such as secret keys and digital certificates, must also be secured at all communication parties and over all wireless links. Due to a number of unique characteristics of wireless networks, designing security algorithms and implementing security mechanisms and protocols in such networks is very difficult. We hope that the following chapters of this book will provide audience with useful information from current research.

References

[1] Wireless at Wikipedia. Retrieved from http://en.wikipedia.org/wiki/Wireless. Accessed 10 November, 2011.

[2] 3G (3rd Generation Mobile Telecommunications) at Wikipedia. Retrieved from http://en.wikipedia.org/wiki/3G. Accessed 10 November, 2011.

[3] Internet of Things at Wikipedia. Retrieved from http://en.wikipedia.org/wiki/Internet_of_Things. Accessed 10 November, 2011.

Chapter 2
Security in Cellular Networks and Communications

Chuan-Kun Wu[1]

Abstract

Cellular Communication has become more and more important in our daily life. The objective of cellular communications has changed from mainly for voice communications as in many years ago, to that mainly for data transmission. The terminal devices for cellular communications also have many more functions other than the functionality for voice communication. Today most cellphones are also personal data assistances (PDAs) as well. Some advanced cellphones are like computers having many applications that used to be for computers. For example, they are able to access the Internet, through which users can conduct a variety of Internet transactions, download and upload data, enjoy on-line entertainment. This is particularly the case in 3G and later generation of networks which are targeted at high speed and wide bandwidth wireless communications. In order to enable sophisticated functionalities in a cellular phone terminal, an operating system is often needed. While a modern and future model of cellphone can give a lot of convenient services to our daily life, it also introduces many security threats, not only threatening the cellphone terminals, but also the cellular communications. This chapter tends to give a primary introduction of common security techniques in cellular communication networks. It is hard to predict what kind of security threats can be encountered in the future.

1 State Key Laboratory of Information Security, Institute of Information Engineering, Chinese Academy of Sciences, Beijing 100093, China, E-mail: ckeu@iie.ac.cn.

Key Terms:

Cellular communications, Universal Mobile Telecommunications System, Authentication and Key Agreement (AKA), authentication vector, privacy protection.

2.1 Introduction

Today people are living in two worlds, a real world and a virtual world. The real world is getting virtually smaller due to the development of transport systems (cars, railways, and aircrafts), and the virtual world is also getting smaller due to the development of wireless communications, which enable people to be connected anywhere, anytime. One of the devices that make most of the contribution to the situation is the kind of cellphones, and the number of cellphones being used today has become very large, which is still growing[1]. Behind the cellphones which are terminal devices of wireless communications, it is the cellular communication systems and perhaps the Internet that connect people together.

A cellular network is a radio network with many fixed-location transceivers known as base stations distributed over land areas. Since the signal of each of the base stations covers only a limited area, there must be sufficient number of base stations in order to have a good signal coverage. In open land areas (without buildings), experiments show that hexagonal distribution of the base stations can achieve a good signal coverage with relatively fewer number of such base stations than other distributions. These hexagons look like cells geometrically (see Fig. 2.1), and hence such a network is called a

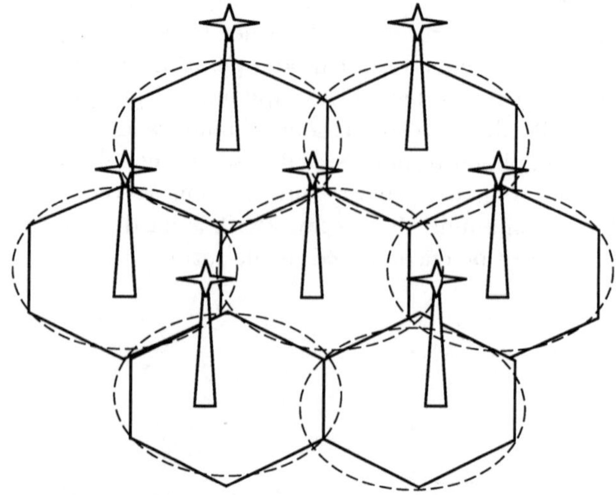

Fig. 2.1 Geometrical view of a cellular network.

cellular network.

One notable advantage of the cellular networks over the traditional ones is the mobility for the network users. With cellular networks, users with a mobile device (transreceiver) can move while the communication still holds. Since there is an overlap of the signals, the transreceivers are able to detect the signal of another base station and switch to the new base station before the signal from the connected base station vanishes. Technically when a mobile user moves from the coverage of one base station to another, the signal of the approaching station becomes stronger, while that of the previous serving base station becomes weaker. At certain stage, there is a signal switch from one base station to another, which may not even give any interruption to the transreceiver or even noticeable to the user. This process is called handover. It is also noted that with cellular networks, signal frequency can be reused, provided that neighboring cells use different ranges of signal frequencies.

The first generation of cellular communication was mostly for the purpose of wireless voice communications and small amount of text such as beeper, and the communication technique used was analogue signals. Just like the wired communications, no or little security techniques regarding the voice communication were considered at that time, because voice communication was treated as having little to do with commercial or other confidential information leakage.

The second generation (2G) of cellular communication systems emerged in the 1990's, primarily using the GSM (Global System for Mobile Communications) standard. There is a substantial difference between the first generation analog cellular communications and the second generation digital cellular communications. The use of digital technique enables many features that were not available in the first generation of cellular networks, including the following: (1) the communication has more robustness against noises, and the quality of voice can be ensured even for long distance communications by using the technique of error-correcting codes; (2) authentication and encryption services are available in the second generation cellular networks which ensure that the wireless communication is secure against "wire-tapping"; (3) short text service as a low-cost service has attracted much interest particularly from young people. Due to the above features of the second generation of cellular networks, and due to the advancement of electric and electronic technologies which make the cellphone terminals getting more and more handy, fancy, and with more and more of other applications. The number of users for the second generation cellular communications has grown to be very large. This figure is still growing, and it is reported that 300 million to 500 million new users are added to the total number of GSM users in the world each year in recent years.

The GSM employs some cryptographic techniques for user authentication and for data encryption. There have been some security weakness for the cryptographic algorithms and the authentication protocols revealed. Due to the increasing demand on data transmission with wireless networks, broadband wireless communications with broad applications are being developed, which leads to the 3rd generation cellular networks (3G) and the long term evolution (LTE) networks. Each of the new generation of cellular networks involve stronger security mechanisms and more sophisticated services.

2.2 Security architecture of cellular communication networks

2.2.1 The first generation of cellular communication networks

Because the first generation of cellular communications used analogue signal which is difficult to provide security services, and at the time when the first generation of cellular communication was in use, the security requirement was not so high, hence the security issues in cellular communications have been addressed only from the second generation of cellular communications with digitalized implementations. So, there is no security provision in the first generation of cellular communication networks.

In fact, in an analogue wireless communication system, an attacker could easily eavesdrop the communication of a cellular phone. A simple radio receiver that can cover the signal frequency for cellular communications can make the eavesdropping easily. There is no confidentiality of the communication data (voice). Moreover, it was technically not too difficult for an attacker to wiretap the identity of a cellphone, so that it is able to make a duplication of the cellphone, and then redirect all the call charges made from the duplicate phone to the owner of the original cellphone. Due to the small scale of the network and small number of cellphone users at that time, these kinds of attacks were not found to be serious threats. The demand for moving from the first generation of analogue communication networks to the second generation of digital ones is not only due to the security concern, but mostly due to the need of more digitalized services, such as text communication and other kinds of digital data exchanges.

2.2.2 The second generation of cellular communication networks

From now on, when we talk about the architecture of cellular communication networks, we mean digitalized communication networks, and they have to be second generation or a later generation of cellular communication networks unless specified otherwise.

Since today most cellular network users are mobile phone ones, where the mobile services cover far more than voice communications, the cellular communication networks are often called mobile networks, and the cellphones are often called mobile phones. Without confusion, in this chapter, by mobile phones and mobile networks we mean the cellphones and the cellular networks respectively.

In cellular communication networks, the transreceivers are also called user equipments (UEs), which are typically identified by a subscriber identity module (SIM, commonly known as SIM card), in combination with a cellphone (or a mobile phone, or a mobile device). A SIM card provides a tamper-proof environment for holding some secret information and execution of some security algorithms. The service providers, known as cellular network providers or mobile network providers, can be identified by two components, the home location register (HLR), which has an authentication center, and a visitor location register (VLR), which is composed of a collection of base stations. The HLR is responsible for issuing each mobile user a unique identity (ID), known as international mobile subscriber identity (IMSI), and a shared secret key. The IMSI and the secret key will be used for the subscriber to authenticate itself to the network. All this information is held by the authentication center in a secure database. On the user side, the information is kept in the SIM card. When a user tries to use the communication services during roaming, the user tries to reach a nearby base station first by sending the ID of the mobile user. The base station collects the information from the user, sends it to a processing unit, the VLR, which then communicates with the network authentication center residing in the HLR to authenticate the user. There can be a long distance communication between a base station and the authentication center, since this part of communication can go through wired networks, or even specific wires owned or hired by the network providers, where strong security techniques can be applied. The security threats during this part of communication are not a big concern for public research. Therefore, the security concerns in cellular communication networks are mainly in the air interface from a mobile user to a nearby base station.

In general, when talking about the security of cellular communications, the wired part of communication is treated as sufficiently secure, the authentication center is treated as being trustworthy, and the SIM card is treated as a tamper-proof hardware device. Although these assumptions are not unconditionally true, since chances for these assumptions to become false is very small, the assumption is widely acceptable. Since cellular communication networks are mostly for the use of mobile communications, we will also alternatively call them mobile communication networks in this chapter. It is noted that there are other kinds of mobile networks such as vehicular ad hoc networks, but this chapter only concerns with the cellular networks for mobile phone communications.

2.2.3 The third generation of cellular communication networks

The second generation of cellular communication networks seems to be able to provide most of the services we need. However, the problem concerned is not just about what kind of services are available. It is also about the quality of services. The prominent improvements of the 3rd generation of cellular communication networks over the 2nd generation ones include the improved security architecture (mutual authentication versus one-way authentication), improved security algorithms, and different radio frequency ranges providing larger communication bandwidth.

In a third generation of cellular communication network, there are three essential network components: the user equipment (UE), a home subscriber server (HSS) which has similar functionalities as an HLR in a 2G network, and many mobility management entities (MME), which have similar functionalities as the VLRs in a 2G network. The authentication process is also very similar to that of 2G networks, but a mutual authentication is enabled in 3G networks. This means that a network has to authenticate itself to the mobile users, apart from the users needing to authenticate themselves to the network.

2.2.4 The 3+ generation of wireless communication networks

The 3+ generation of wireless communication networks including long term evolution (LTE) networks (also known as 3.5G), WiMax (4G), WiFi (4G). Apart from LTE which is based on 3G network and hence has very similar architecture with that of 3G networks, other networks such as WiMax and WiFi have very different architectures and are not appropriate to be called cellular networks, and in this case their security problems are beyond the scope of this chapter and are not considered.

2.3 Security techniques in GSM networks

It is noted that the security threats become more serious with the scale of applications increases, i.e., under the same environment, a small scale application encounters fewer attacks than a larger scale application in the same environment. This is reasonable because on one hand, a larger scale application attracts more interests including those from attackers, hence more attacks may occur. On the other hand, launching an attack often involves some cost, so that the scale of an application is also a reason concerned with whether it is worth for the potential attackers to launch an attack.

It is known that the number of GSM users is far larger than that of those using other communication networks following the first generation in analogue signals, we will mainly focus on the security of GSM networks in this section.

2.3.1 User authentication in GSM

In GSM system, all the mobile users have to authenticate themselves to the network before the network can provide services. In the GSM networks, a mobile user denoted as user equipment (UE) which includes a compatible mobile device (e.g. a cell phone) and a SIM card, get their unique ID from a mobile network authentication center, which is located in or collaboratively working within a home location register (HLR). When a mobile user connects to the network, a nearby base station is contacted, which is managed by a visitor location register (VLR, often multiple base stations are being managed by a same VLR). The Authentication and Key Agreement (AKA) protocol for a GSM network can be depicted in Fig. 2.2.

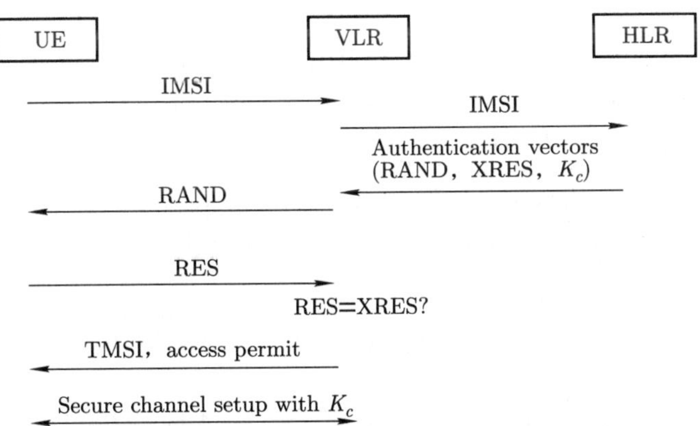

Fig. 2.2 Authentication and key agreement (AKA) process in GSM.

From Fig. 2.2, it can be seen that, when a mobile user tries to connect to the mobile networks, it first sends its IMSI serving as its identity to a nearby base station, which collects information and transfers to the VLR. The VLR is able to find which HLR the IMSI user belongs to, and sends the IMSI to the corresponding HLR. The HLR also serves as an authentication center and after checking the user as valid, creates a number of authentication vectors, and sends the authentication vectors to the VLR. In each of the authentication vectors, there are three components, they are a 128-bit random number RAND, a 32-bit expected response XRES=A3 (K_i, RAND), and a 64-bit data encryption key K_c=A8 (K_i, RAND), where A3 and A8 are two standard encryption algorithms in the GSM system, and K_i is the user key shared by the mobile user and the authentication center. When the VLR receives the authentication vectors, it chooses one of the authentication vectors, sends the RAND in the authentication vector to the mobile user. When the mobile user receives the RAND, its SIM card has the user key K_i and the encryption algorithms A3 and A8, and is able to create RES=A3 (K_i, RAND), and sends RES back to the VLR. The VLR then compares the RES received from the mobile user with the XRES as in the authentication vector received from the HLR. If the user is a valid one, then the equality of RES=XRES should hold, and hence the authentication is passed. In this case, the VLR generates a temporary mobile subscriber identity (TMSI), and has setup a secure communication channel with the mobile user under the protection of a common session key K_c, because the mobile user is also able to compute K_c. Then the data transmission between the mobile user and the contacting base station will be protected by an encryption algorithm named as A5. The use of TMSI is to provide privacy protection of the IMSI to certain degree. This issue will be further discussed later.

2.3.2 The authentication algorithms A3 and A8

The algorithms of A3 used for user/equipment authentication and the algorithm A8 used for generating a session key K_c are all based on a cryptographic algorithm named as COMP128, while the algorithm A5 used for data encryption is a stream cipher. The COMP128 is a keyed hash function that takes a 128-bit key and a 128-bit random number as input, and generates a 96-bit hash code. The 96-bit output then is split into a 32-bit XRES and a 64-bit K_c, as shown in Fig. 2.3.

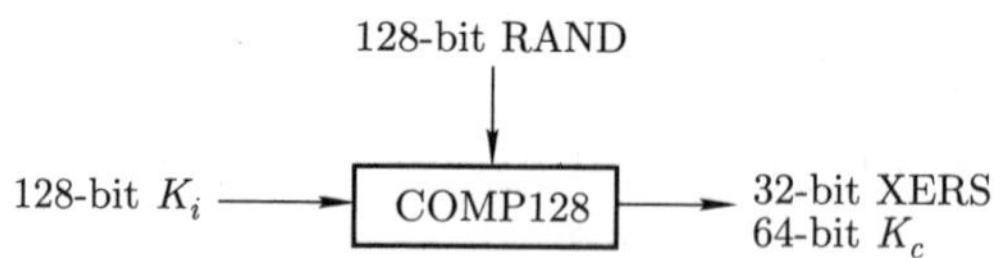

Fig. 2.3 The generation of XRES and K_c using COMP128.

The COMP128 works as follows: first it loads a 128-bit key and a 128-bit RAND by concatenation into a 32-byte array, and then a compression function is called for 5 times, mainly functioning at the positions where the 128-bit key is loaded.

The algorithm COMP128 was meant to be an industry standard yet to remain secret to the public. However with partial information being accidentally released via the Internet, the Smartcard Developer Association (SDA) and two U.C. Berkeley researchers, Ian Goldberg and David Wagner, jointly broke the COMP128 algorithm which leads the cloning of SIM cards possible. They demonstrated that the A8 algorithm takes a 64-bit key, but ten key bits were set to zero. The attack on the A8 algorithm demonstrated by Goldberg and Wagner takes just 2^{19} queries to the GSM SIM, which takes roughly 8 hours. Later it was shown by Josyula et al. of IBM that COMP128 can be broken in less than a minute.

Noticed that, even though the COMP128 algorithm was shown to be insecure, the GSM system had to provide services for even increasing number of users, and hence the algorithm cannot be abandoned. In order to provide a better security for later GSM users, some modifications for the COMP128 were made, which leaded to COMP128 version 2 and version 3. The broken of COMP128 algorithm (version 1) is a typical example to show that "security by obscurity" is not a good practice to provide security. The COMP128-2 and COMP128-3 are also secret algorithms which have not been subject to cryptanalysis. COMP128-3 fixes the problem of COMP128-1 where 10 bits of the Session Key (K_c) were set to zero.

It should be noted that the algorithms A3 and A5 are used within a specific network and their subscribers. In a different network, the A3 and A5 algorithm can well be different. So there is no need for global standardization of them, although they have been globally standardized. A good effect of using non-standard algorithms is that upgrading of the algorithms can be done relatively easily, and security flaws revealed in one network may not be a threat to another network due to the use of a different set of algorithms.

2.3.3 The data encryption algorithms A5

In the GSM system, the A5 algorithm is very different from the A3 and A8 algorithms. It is a stream cipher used to provide confidentiality for messages in over-the-air transmission. It is different also in the sense that the A5 algorithm has to be standardized globally so that services during worldwide roaming can be provided.

The A5 algorithm also has three versions, where the first two versions were meant to be kept secret from the public. However the general design was leaked in 1994, and the algorithms were entirely reverse engineered in 1999 by Marc Briceno, and security analysis on the algorithms became public since then. The first version of the algorithm, named as A5/1 or the original A5 algorithm was developed in 1987 mainly for GSM users in Europe. However for the purposes of export, the second version of the algorithm, named as A5/2, was developed with the security being deliberately weakened, and was used in the United States. It is probably to the surprise of the designers how weak the A5/2 algorithm is, which could even affect the security of the systems using the A5/1 algorithm by default.

The A5/1 stream cipher algorithm uses three linear feedback shift registers (LFSRs) over the binary field GF(2) of lengths 19, 22 and 23 respectively, with feedback polynomials being $x^{19}+x^{18}+x^{17}+x^{14}+1$, $x^{22}+x^{21}+1$ and $x^{23}+x^{22}+x^{21}+x^{8}+1$ respectively. All the three LFSRs clock irregularly in a stop/go fashion. More precisely, there is a clocking bit for each of the three registers. At each clock cycle, a register is clocked unless its clocking bit does not agree with either of the clocking bits for the other two registers. This non-clocking happens when the clocking bit of the current register is 1 (0) when the other two clocking bits are both 0 (1), and happens with probability 1/4. The final output bit of the A5/1 algorithm is the exclusive-or of the output bits of the three registers.

2.3.4 The security weakness of the algorithms A5

A number of different attacks have been found soon after the A5/1 algorithm being revealed to the public, while the attacks on the A5/2 algorithm are more efficient. In 1999, Ian Goldberg[2] cryptanalyzed A5/2 in the same month when it was made public, and showed that it was so weak that can be broken in real time.

In 2003, Barkan et al.[3], and Ekdahl and Johansson[4] published their attacks on A5/1 algorithm, and a more efficient attack was given in[5] which can break A5/1 in real time, or at any later time. In fact, Barkan's approach was not directly to attack the A5/1 algorithm, it is to use the efficient attack on A5/2 algorithm to trigger the encryption key used by the A5/1 algorithm since both of the algorithms are assumed to use a same encryption key. Barkan's attack can be depicted in Fig. 2.4.

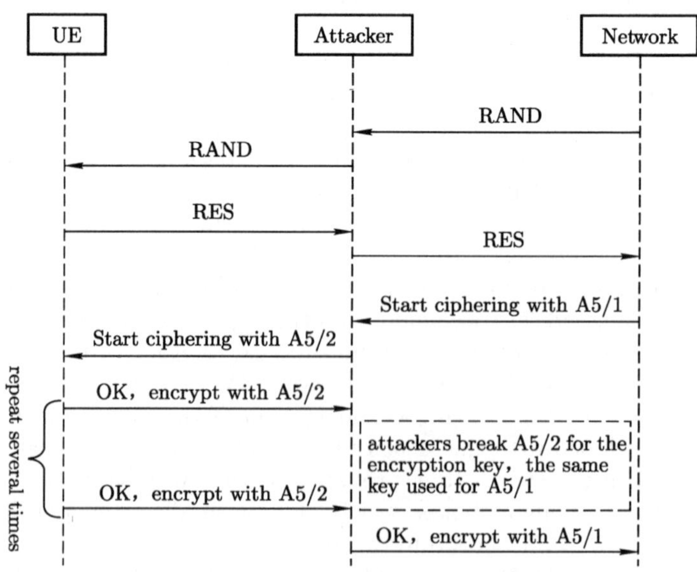

Fig. 2.4 Attacking on A5/1 via attacking on A5/2.

Alternatively, the attacker can record the RAND, the RES, and the immediate conversation, and later on use the same RAND and RES to forge another AKA process with the network claiming to use the A5/2 algorithm.

Once the A5/2 is broken, the key must be the same as what was used in the previous recorded conversation so that it can be decrypted.

2.3.5 The algorithms A5/3: a complete new version

To overcome the weakness of the A5/1 algorithm (the A5/2 was abandoned due to its security being too weak), the third version of the algorithm, named as A5/3, was introduced in the GSM system. The A5/3 changed the philosophy of "security by obscurity". Instead it used an algorithm KASUMI[6], with some small modifications for easier hardware implementation and to meet other requirements for 3G mobile communications security. A5/3 is a strong encryption algorithm created as part of the 3rd Generation Partnership Project (3GPP).

Different from A5/1 and A5/2, the KASUMI algorithm used by A5/3 is a block cipher. The KASUMI algorithm can be described briefly as follows: given a 128-bit key and a 64-bit message as its inputs, the message is split into a 32-bit left half and a 32-bit right half, denoted as $m = L_0||R_0$. Then the KASUMI processes the message in 8 rounds of iteration, and finally outputs $L_8||R_8$.

As in many other block ciphers, an initial key is used to produce many round-keys. In KASUMI, each round of encryption needs three round-keys KL_i, KO_i and KI_i. Write the initial 128-bit key as

$$K = K_1||K_2||K_3||K_4||K_5||K_6||K_7||K_8,$$

each K_i is a 16-bit string. Define

$$K' = K \oplus C = K'_1||K'_2||K'_3||K'_4||K'_5||K'_6||K'_7||K'_8,$$

where $C = 0 \times 123456789\text{ABCDEFFEDCBA}9876543210$. Then the round-keys are as follows:

$$\begin{aligned}
KL_{i,1} &= K_i \lll 1 \\
KL_{i,2} &= K'_{i+2} \\
KO_{i,1} &= K_{i+1} \lll 5 \\
KO_{i,2} &= K_{i+5} \lll 8 \\
KO_{i,3} &= K_{i+6} \lll 13 \\
KI_{i,1} &= K'_{i+4} \\
KI_{i,2} &= K'_{i+3} \\
KI_{i,3} &= K'_{i+7}
\end{aligned}$$

where "$X \lll j$" means cyclic shift of X to the left by j bits, and the subscript index should take the modulo 8 value.

There are three core functions in the KASUMI, they are named as FL, FO, and FI. The function FL in round i, denoted as FL_i, is defined as taking the 32-bit round-key $KL_i = KL_{i,1}||KL_{i,2}$ and a 32-bit data $I = L||R$ as inputs, and produces a 32-bit output. More precisely, first do the bitwise AND operation for L and $KL_{i,1}$, then the result performs a cyclic shift to the left by one bit, then the result is XOR'ed with R to get the right half of the output R': $R' = ((L \wedge KL_{i,1}) \lll 1) \oplus R$. Then do the bitwise OR operation for R' and $KL_{i,2}$, perform a cyclic shift of the result to the left by one bit, and then the result is XOR'ed with L to get the left half of the output $L' = ((R' \wedge KL_{i,2}) \lll 1) \oplus L$. Finally the output of the function $FL_i(KL_i, I)$ is $I' = L'||R'$.

The function FO in round i, denoted as FO_i, is defined as taking as input a 32-bit data I, a 48-bit round-key $KO_i = KO_{i,1}||KO_{i,2}||KO_{i,3}$ and a 48-bit round-key $KI_i = KI_{i,1}||KI_{i,2}||KI_{i,3}$, and produces a 32-bit output. More precisely, denote $I = L_0||R_0$, for $j = 1, 2, 3$, perform the following operations:

$$\begin{cases} R_j = FI_{i,j}(L_{j-1} \oplus KO_{i,j}, KI_{i,j}) \oplus R_{j-1} \\ L_j = R_{j-1} \end{cases}$$

Then the output of the FO_i function is the 32-bit data block $L_3||R_3$.

It is noted that the computation of the function FO_i involves another $FI_{i,j}$ function. For a given i and j, where $1 \leqslant i \leqslant 8$ and $1 \leqslant j \leqslant 3$, an FI-function $FI_{i,j}$ takes a 16-bit data x and a 16-bit subkey $KI_{i,j}$ as input, and produces a 16-bit data. More precisely, the data x is split into a 9-bit left part l_0 and a 7-bit right part r_0, similarly the subkey $KI_{i,j}$ is also split into a 9-bit left part and a 7-bit right part as $KI_{i,j} = KI_{i,j,1}||KI_{i,j,2}$. There are also two S-boxes needed, one is S_9 that maps a 9-bit input into a 9-bit output, and the other is S_7 that maps a 7-bit input into a 7-bit output. The detailed definition of the S-boxes is not specified here, as they are widely and publicly available. Denote $LS_7(y)$ as the least significant 7-bit part of y. Then the function $FI_{i,j}$ is defined by the following series of operations:

$$\begin{array}{ll} l_1 = r_0 & r_1 = S_9(l_0) \oplus (00||r_0) \\ l_2 = r_1 \oplus KI_{i,j,2} & r_2 = S_7(l_1) \oplus LS_7(r_1) \oplus KI_{i,j,2} \\ l_3 = r_2 & r_3 = S_9(l_2) \oplus (00||r_2) \\ l_4 = S_7(l_3) \oplus LS_7(r_3) & r_4 = r_3 \end{array}$$

The output of $FI_{i,j}$ is the 16-bit data block $l_4||r_4$.

With the introduction of the core functions in KASUMI, it is easy to introduce the encryption process. Given a 128-bit key K and a 64-bit input message m, the message is first split into two 32-bit halves, $m = L_0||R_0$. Then for each round i with $1 \leqslant i \leqslant 8$, the operation of KASUMI on the i-th round is as follows:

$$R_i = L_{i-1} \quad L_i = R_{i-1} \oplus f_i(L_{i-1}, RK_i)$$

where RK_i is the round key which in fact is defined as a triplet of subkeys (KL_i, KO_i, KI_i). The function f_i differs in odd rounds and even rounds. For round number $i = 1, 3, 5, 7$, the f-function is defined as:

$$f_i(L_{i-1}, RK_i) = FO_i(FL_i(L_{i-1}, KL_i), KO_i, KL_i)$$

and for round number $i = 2, 4, 6, 8$, the f-function is defined as:

$$f_i(L_{i-1}, RK_i) = FL_i(FO_i(L_{i-1}, KO_i, KI_i), KL_i)$$

The final output of the algorithm is the 64-bit data block $L_8||R_8$.

Although KASUMI is a minor modification of MISTY suitable for hardware implementation, it is surprise to note that, in 2010, Dunkelman et al. published a paper[7] claiming that they could break KASUMI with a related key attack and very modest computational resources. Interestingly, the attack is ineffective against MISTY.

2.3.6 The inherent security weakness of 2G networks

It is not known how far the A5/3 algorithm can secure the GSM system in the sense of air data confidentiality, and given the weakness of A5/3 having been found[8], a new algorithm A5/4 may be in place in the near future. However, the inherent AKA process of GSM system has some fatal weakness. More precisely it only provides one-way authentication. i.e., it enables the users to authenticate themselves to the network, and does not provide functionality for the network to authenticate itself to the end users. This may cause some attacks by false base stations. Fig. 2.5 depicts how a false base station could eavesdrop a victim mobile user.

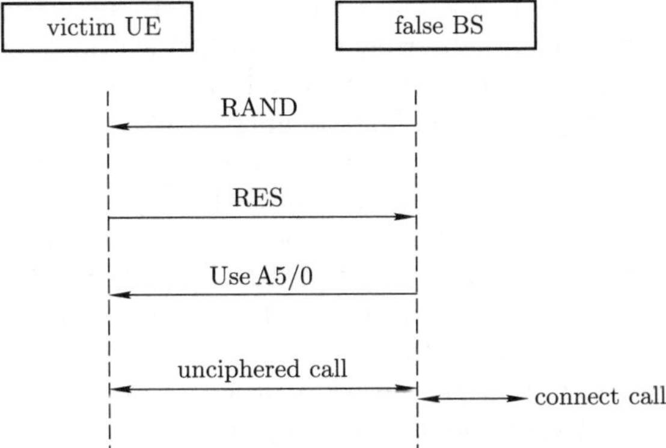

Fig. 2.5 Eavesdropping by a false base station.

2.4 Security techniques in 3G networks

The technology for mobile communication evolves quickly, especially in recent years. The initial purpose of cellular communication networks were meant to serve for voice communications, while the second generation of cellular communication systems can also provide short message services and even some extended services such as GPRS. The most significant improvement of the second generation of mobile networks over the first generation is the information security, including the functionality for end user authentication and data confidentiality. However, the lack of mutual authentication of the GSM system suffers the attack by false networks. On the other hand, the GSM system has limited channel bandwidth, which is perhaps enough for voice communications. With the development of mobile networks, the increased functionality, and the demand of mobile devices, a wider wireless bandwidth is needed while a higher security is to be provided. This leads to the 3G mobile communication networks (or 3G networks for short).

There are different techniques in 3G networks, including WCDMA, CDMA2000, and TD-SCDMA networks. There are many core techniques in common, that is code division multiple access (CDMA) techniques, and there are also essential differences between any of the networks. The most similar networks from architecture point of view are WCDMA and CDMA2000. They are also the most widely used 3G networks. Here we will introduce the security architecture of these networks (say WCDMA), and without confusion we simply name it as 3G network security architecture.

2.4.1 The mutual authentication in 3G networks

In a 3G network, the network components are commonly known as user equipment (UE), eNodeB (essentially a base station), a mobility management entity (MME), and a home subscriber server (HSS) who also serves as network authentication server. The authentication and key agreement (AKA) process can be depicted in Fig. 2.6. The functionality of MME is very much like the VLR as in GSM systems, that of HSS is very much like that of HLR as in GSM systems, where the eNodeB is a connection between end user and an MME. To make it simple and comparable with the AKA process in GSM networks, we treat eNodeB as part of MME when dealing with the security functionalities. The process of authentication and key agreement in a 3G network can be depicted in Fig. 2.6.

It is seen from Fig. 2.6 that the AKA process in 3G networks are almost the same as that in 2G networks, except that in 3G networks, the authentication vectors are 5-tuples versus triplicates as in 2G networks, and there is an AUTH send from the network to the end user for verifying the network authenticity. The temporary user equipment identity in 3G networks is named

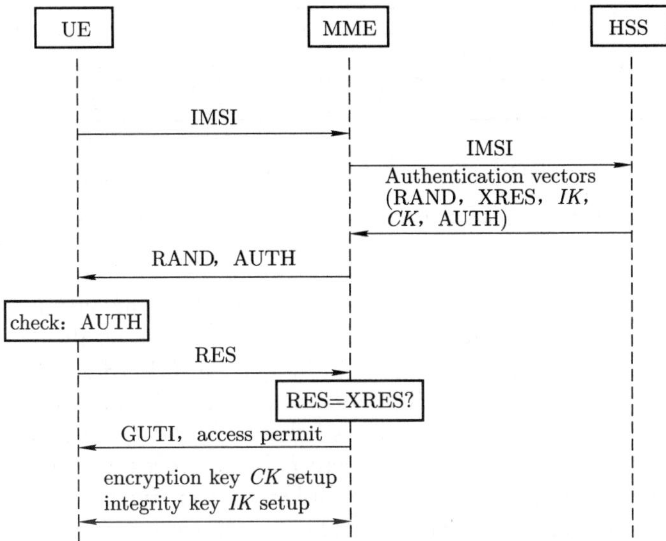

Fig. 2.6 The AKA process in a 3G network.

as globally unique temporary identity (GUTI).

2.4.2 The confidentiality algorithm f_8 and the integrity algorithm f_9

Although there are some similarities of the AKA processes in different networks, there are substantial differences as well. In 3G networks, the 5-tuple authentication vectors are generated by algorithms totally different from that in 2G networks. More precisely, there are 11 security algorithms defined in 3G networks, they are f_0, f_1^*, $f_1 \sim f_9$, where f_0 is a pseudorandom number generator that generates random challenges, f_1 is used to generate a message authentication code (MAC) to be part of the authentication token AUTH, f_1^* is used for the resynchronization of message authentication, f_2 is used to generate the expected response (XRES) corresponding to the challenge RAND, f_3 is used to generate an encryption key CK, f_4 is used to generate an integrity key IK, f_5 is used to generate an anonymity key AK. The functions $f_1 \sim f_5$ are responsible for generating the authentication vectors, and Fig. 2.7 shows how they work in general.

Note from Fig. 2.7 that the inputs of f_1 also includes AMF and SQN, they are authentication management field (AMF) and sequence number (SQN), two parameters known to both the end user and the home subscriber server. The common inputs to all the five functions are RAND and K, where K is the long term user key shared between the mobile user (in a SIM or USIM card)

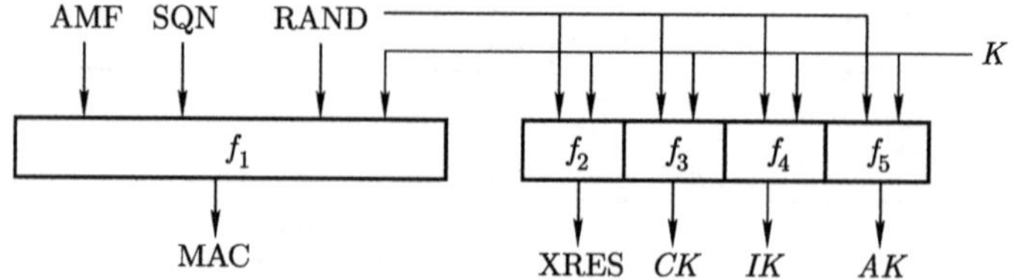

Fig. 2.7 Authentication vector generation in 3G networks.

and the network authentication center resides in HSS. The output of f_1 is not yet the authentication token AUTH, which in fact can easily be computed given the inputs and outputs of $f_1 \sim f_5$. In fact, AUTH=(SQN$\oplus AK$)||AMF|| MAC, where $\oplus$ means bitwise XOR operation and || is concatenation.

The other functions, $f_6 \sim f_9$, are as follows: f_6 and f_7 are used to provide enhanced user identity encryption, where f_6 is the process for encryption, and f_7 is the inverse of f_6. f_8 is a stream cipher that encrypts the user-network air-interface communication after mutual authentication is successful, and f_9 is an algorithm for generating a message authentication code (MAC) for the signaling messages. Fig. 2.8 shows the structure of algorithms f_8 and f_9.[9]

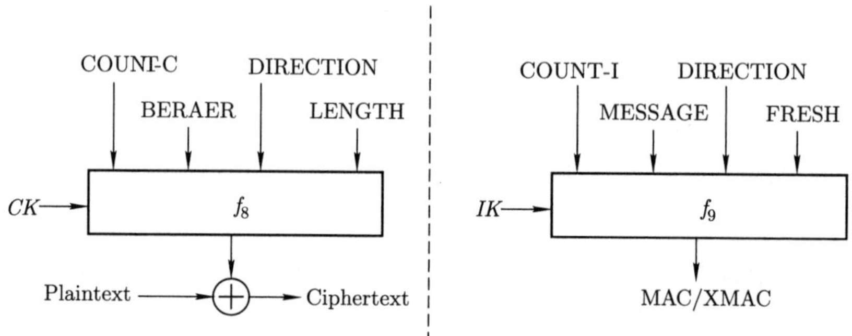

Fig. 2.8 The structure of f_8 and f_9.

Although there are 11 functions in the 3G networks, apart from the supplementary security functions such as f_0, f_1^*, f_6 and f_7, most security functions take KASUMI and AES as the core algorithm.[10,11] Although f_8 is a stream cipher, and KASUMI is a block cipher, it uses the output feedback (OFB) mode of KASUMI to build a stream cipher from a block cipher. As has been pointed out in section 3, since the algorithm KASUMI has some security problems revealed, the related functions that are built upon the KASUMI algorithm may also have security problems[12]. Fortunately, this situation is improved in the LTE networks.

2.5 Security techniques in LTE networks

The 3GPP organization has been working on the quality and capability of mobile communications. With some limitations of the 3G networks emerged, including the security limitations, a new generation of networks named long term evolution (LTE) is proposed. The LTE networks are targeted at an even higher rate data transmission than that of 3G networks and hence can provide more services. Naturally with increased number and intensity of services, the security becomes more sensitive. Given that the 3G networks use KASUMI as the core cryptographic algorithm on top of which many security functions are built, and the KASUMI algorithm has some security problems revealed. The LTE networks tend to employ different suits of algorithms, named as EEA and EIA, mainly for providing confidentiality (the function f_8) and integrity (the function f_9) services respectively. Since many of the security functions (e.g. $f_1 \sim f_7$) can be different from network to network, some internal modification within a network operator is practically possible. However, the security functions for data confidentiality and integrity have to be globally standardized and their security is of great interest and is also the most concerned.

2.5.1 The confidentiality and integrity algorithm sets for LTE

The first suite of the algorithms, 128-EEA1/128-EIA1, is based on a stream cipher called SNOW-3G, designed by the Security Algorithms Group of Experts (SAGE), part of the European standards body ETSI. The second suite of the algorithms, 128-EEA2/128-EIA2, is based on the Advanced Encryption Standard (AES). The third suite of the algorithms, 128-EEA3/128-EIA3, is based on a newly proposed stream cipher named ZUC. Due to the well availability of the AES algorithm discussions, we will only look into the algorithms SNOW-3G and ZUC, yet in a very brief manner, which intend to show their structural similarities and differences. First, we give a diagram to show the structure of SNOW-3G (see Fig. 2.9).

From Fig. 2.9, it is seen that SNOW-3G has two essential parts[13], a linear feedback shift register (LFSR) of order 16 defined over the finite field $GF(2^{32})$, where α is a specific field element, and a finite state machine (FSM) composed of three memory registers R_1, R_2 and R_3, and two S-boxes, S_1 and S_2. The output z is a sequence of elements over $GF(2^{32})$. i.e., each output of the algorithms is a 32-bit data block.

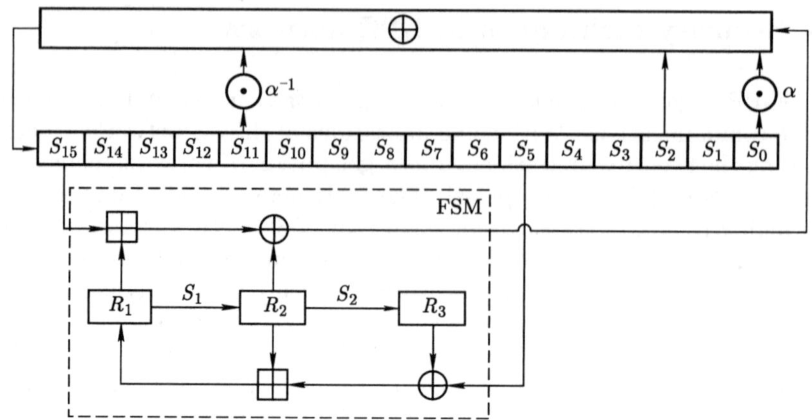

Fig. 2.9 The structure of SNOW-3G.

2.5.2 A new stream cipher ZUC

The name of the stream cipher ZUC is after a famous Chinese mathematician in the history named as Zu Chongzhi. The structure of ZUC, as shown in Fig. 2.10, looks to have some similarities with that of SNOW-3G. In fact, they are similar in a few phases. First, both of the algorithms have two components: an LFSR and a finite state machine. Second, they both use 128-bit seed key and output 32-bit key streams. Third, they both employ a mixture of different operations, e.g. addition and multiplication over a finite field, exclusive-OR, and addition modular an integer. However, it is also easy to find some substantial differences. First, ZUC used an LFSR defined over a prime finite field GF$(2^{31}-1)$ which seems to have more complicated algebraic structure when being viewed over the binary field GF(2). Second, ZUC has a bit-reorganization operation which breaks the algebraic structure of the contents from the LFSR cells. Third, the finite state machine component in ZUC seems to be more complicated than that in SNOW-3G. And fourth, ZUC has more operations than those used in SNOW-3G. It is not surprising that the performance of ZUC is slightly degraded compared with that of SNOW-3G. On the other hand, the structural differences of ZUC compared with SNOW-3G make the two algorithms not likely to stand or fall together, as has been pointed by the two review reports[19,20].

It should be pointed out that the security functions f_8 and f_9 as in LTE networks are not bounded by the use of SNOW-3G or ZUC. SNOW-3G is a standard algorithm and ZUC is in the process of becoming a standard algorithm. With the development of mobile networks, there may be more cryptographic algorithms introduced in the future, while some of the existing algorithms may be abandoned.

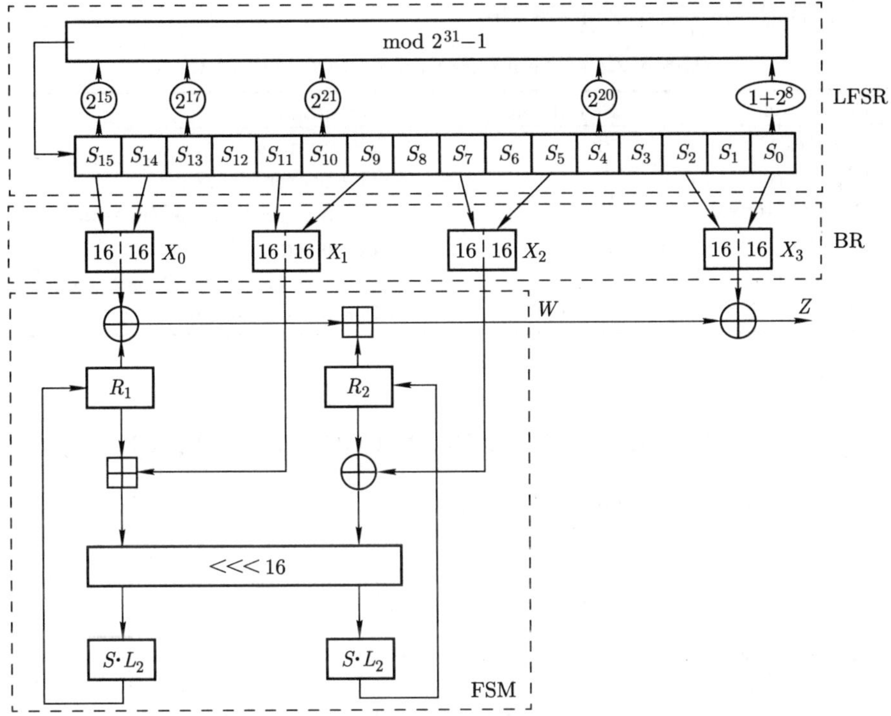

Fig. 2.10 The structure of ZUC.

2.5.3 The confidentiality/integrity algorithm set 128-EEA3/128-EIA3

Within the security architecture of the LTE system, there are standardized algorithms for confidentiality and integrity. Two sets of algorithms 128-EEA1/128-EIA1 and 128-EEA2/128-EIA2 have been specified as standard[9,10]. The third set of algorithms 128-EEA3/128-EIA3 is based on the stream cipher ZUC as described above.

The confidentiality algorithm 128-EEA3 is a stream cipher that is used to encrypt/decrypt blocks of data using a confidentiality key CK. The block of data may be between 1 and 20 000 bits long. The integrity algorithm 128-EIA3 computes a 32-bit Message Authentication Code (MAC) of a given input message using an integrity key IK. Since ZUC is an algorithm needing an initialization vector (IV) as well as an initial key for the initialization, there will be an IV involved both in the confidentiality algorithm 128-EEA3 and the integrity algorithm 128-EIA3[14,15].

The inputs to the algorithms 128-EEA3/128-EIA3 are given in Table 2.1. The encryption key for data confidentiality is CK and that for data integrity is IK. Both are a string of 128 bits.

Table 2.1 The inputs to 128-EIA3/128-EIA3

Parameter	Size in bits	Meaning
COUNT	32	The counter
BEARER	5	The bearer identity
DIRECTION	1	The direction of transmission
CK/IK	128	The integrity key
LENGTH	32	The bits of the input message
M	LENGTH	The input message

Let

$$\text{COUNT=COUNT}[0]\|\text{COUNT}[1]\|\text{COUNT}[2]\|\text{COUNT}[3]$$

be the 32-bit counter, where COUNT[i] ($0 \leqslant i \leqslant 3$) are bytes. The 128-bit initialization vector for 128-EEA3 is set as

$$\text{IV} = \text{IV}[0]\|\text{IV}[1]\|\text{IV}[2]\|\ldots\|\text{IV}[15],$$

where IV[i] ($0 \leqslant i \leqslant 15$) are bytes, defined by

$$\begin{aligned}
&\text{IV}[0] = \text{COUNT}[0], \quad \text{IV}[1] = \text{COUNT}[1],\\
&\text{IV}[2] = \text{COUNT}[2], \quad \text{IV}[3] = \text{COUNT}[3],\\
&\text{IV}[4] = \text{BEARER}\|\text{DIRECTION}\|00,\\
&\text{IV}[5] = \text{IV}[6] = \text{IV}[7] = 00000000,\\
&\text{IV}[8] = \text{IV}[0], \quad \text{IV}[9] = \text{IV}[1],\\
&\text{IV}[10] = \text{IV}[2], \quad \text{IV}[11] = \text{IV}[3],\\
&\text{IV}[12] = \text{IV}[4], \quad \text{IV}[13] = \text{IV}[5],\\
&\text{IV}[14] = \text{IV}[6], \quad \text{IV}[15] = \text{IV}[7].
\end{aligned}$$

The initialization vector used for 128-EIA3, IV=IV[0]||IV[1]||IV[2]||...|| IV[15], is defined by

$$\begin{aligned}
&\text{IV}[0] = \text{COUNT}[0], \quad \text{IV}[1] = \text{COUNT}[1],\\
&\text{IV}[2] = \text{COUNT}[2], \quad \text{IV}[3] = \text{COUNT}[3],\\
&\text{IV}[4] = \text{BEARER}\|(000)_2, \quad \text{IV}[5] = (00000000)_2,\\
&\text{IV}[6] = (00000000)_2, \quad \text{IV}[7] = (00000000)_2,\\
&\text{IV}[8] = \text{IV}[0] \oplus (\text{DIRECTION} \ll 7), \quad \text{IV}[9] = \text{IV}[1],\\
&\text{IV}[10] = \text{IV}[2], \quad \text{IV}[11] = \text{IV}[3],\\
&\text{IV}[12] = \text{IV}[4], \quad \text{IV}[13] = \text{IV}[5],\\
&\text{IV}[14] = \text{IV}[6] \oplus (\text{DIRECTION} \ll 7), \quad \text{IV}[15] = \text{IV}[7].
\end{aligned}$$

There is not much to say about the 128-EEA3 algorithm, because it is basically to run the ZUC algorithm to encrypt the message using the encryption key CK and the initialization vector IV. It only needs to note that the key is used only to encrypt a message of up to 20 000 bits.

The 128-EIA3 algorithm, however, has something more than the ZUC. More precisely, Let $N = \text{LENGTH} + 64$ and $L = \lceil N/32 \rceil$. Let ZUC generate L 32-bit key words $z[0]$, $z[1]$, $\cdots$, $z[L-1]$ with the initial key IK and the initialization vector IV is defined as above, where $z[0]$ is the first key word generated by ZUC, $z[1]$ is the next, and so on. Let $k[0]$, $k[1]$, $\cdots$, $k[31]$, $k[32]$, $\cdots$, $k[N-1]$ be the key bit stream corresponding to the above key words $z[1]$, $\cdots$, $z[L-1]$. Then $N = 32 * L$.

For each $i = 0, 1, 2, \cdots, N-32$, let $k_i = k[i] \| k[i+1] \| \ldots \| k[i+31]$. Then each k_i is a 32-bit word. LET T be a 32-bit word. Set $T = 0$, and for each $i = 0, 1, 2, \cdots$, LENGTH-1, if $M[i] = 1$, then set $T = T \oplus k_i$. At last let $T = T \oplus k_{\text{LENGTH}}$. Finally we take $T \oplus k_{N-32}$ as the output MAC, i.e. $\text{MAC} = T \oplus k_{N-32}$.

2.5.4 The security flaws and improvements of ZUC

The ZUC algorithm has an initialization process before actual key stream can be produced. The initialization can be depicted in Fig. 2.11.

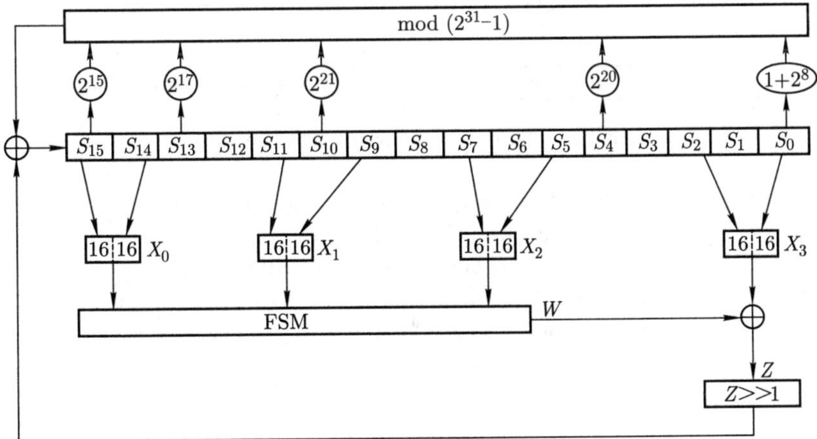

Fig. 2.11 The initialization of ZUC version 1.0.

Certain effort has been made on evaluating the security of the ZUC algorithm. First of all, the designers have made a comprehensive security analysis and evaluation. Then some professional evaluation groups gave their evaluation reports[19,20]. Surprisingly none of these evaluations has found obvious security flaws of ZUC.

It was realized that the ZUC algorithm has some security problems not long after the publication of the algorithm specification. The first flaw was notified by both the designers as well as some external members, and has been reported at the first international workshop on ZUC conducted in December 2010. It was found that the initialization process does not keep key entropy due to that $\boldsymbol{z}$ is involved in updating the LFSR feedback. In fact, the very first version of ZUC was to use $\boldsymbol{w}$ to be involved in updating the LFSR, and the change was to base on the assumption that $\boldsymbol{z}$ is more likely to be balanced than $\boldsymbol{w}$. In fact, a more serious problem was found by Wu[21] that, when $\boldsymbol{z}$ or $\boldsymbol{w}$ was truncated from a 32-bit string into a 31-bit string, it can cause problems. Denote that as $\boldsymbol{z}$, let the feedback of the LFSR be $\boldsymbol{x}$, then when $\boldsymbol{z} = \boldsymbol{x}$ and $\boldsymbol{z} = 2^{32} - \boldsymbol{x}$, these two values will all result in $\boldsymbol{z} \oplus \boldsymbol{x} = \boldsymbol{0}$, and hence cause attacks.

Notifying these security flaws, the version 1.5 of ZUC released on 4th January 2011[23] has some changes on the initialization of ZUC, which can be depicted in Fig. 2.12.

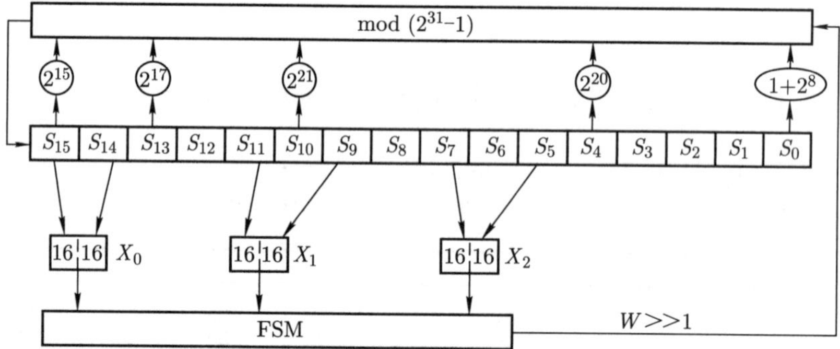

Fig. 2.12 The initialization of ZUC version 1.5.

It is noted that there are two changes in the ZUC version 1.5 over the earlier version. One of the changes is that $\boldsymbol{w}$ instead of $\boldsymbol{z}$ is involved in updating the content of s_{15} in the LFSR, and another change is that the operation on $\boldsymbol{w} \gg 1$ with the feedback of the LFSR is addition modulo $2^{31} - 1$ instead of XOR which was found to cause problems. These two simple changes seem to have solved the security flaws found so far.

2.5.5 The security flaws and an improvement of 128-EIA3

Regardless which version of the ZUC algorithm is used, the 128-EIA3 in its earlier versions were found to be insecure[22]. The process of 128-EIA3 can be depicted in Fig. 2.13.

The attack found by Fuhr et al.[22] on 128-EIA3 before the version 1.5 is as follows: let message M produces a MAC using a key IK and an initialization vector IV. Let $M' = 1||M$ be the concatenation of 1 and the message M.

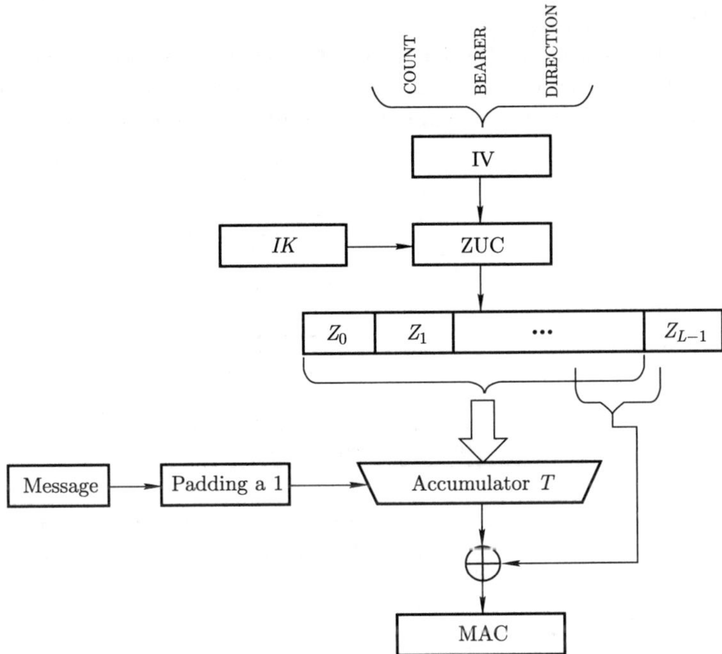

Fig. 2.13 The process of the EIA3 MAC computation.

Then it is easy to verify that the new authentication code MAC′ for M' under the same set of key/IV will have 31 bits in common with MAC, so one can guess the other bit of MAC′ with a probability of 50% to succeed. This high probability of successfully forging an authentication code is not acceptable.

It is noted that the cause of forgery authentication code to be possible was mainly from the observation that the last "mask" to finally producing a MAC was not really random for different messages, particularly for the two messages in the form above. Therefore, a simple change can amend the security flaw. In the version 1.5 of 128-EIA3, the following change has been made:

Let ZUC generate a key stream of $L = \lceil \text{LENGTH}/32 \rceil + 2$ words, each word is 32 bits. Denote the generated bit string by $z[0], z[1], \cdots, z[32\times(L-1)]$, where $z[0]$ is the most significant bit of the first output word of ZUC and $z[31]$ is the least significant bit.

For each $i = 0, 1, 2, \cdots, 32 \times (L-1)$, let $z_i = z[i] \| z[i+1] \| \ldots \| z[i+31]$. Then each z_i is a 32-bit word. Let T a block of 32 bits word. Set $T = 0$.

For cach $i = 0, 1, 2, \cdots$, LENGTH 1, if $M[i] = 1$, thcn

$T = T \oplus z_i$.

Set

$T = T \oplus z_{\text{LENGTH}}$.

Finally we take $T \oplus z_{32\times(L-1)}$ as the output of MAC, i.e.

MAC= $T \oplus z_{32\times(L-1)}$.

In brief, the change of the EIA3 was to use the next word generated by ZUC as the mask, as depicted in Fig. 2.14, where in the earlier version a 32-bit block immediately following the computation of T was used.

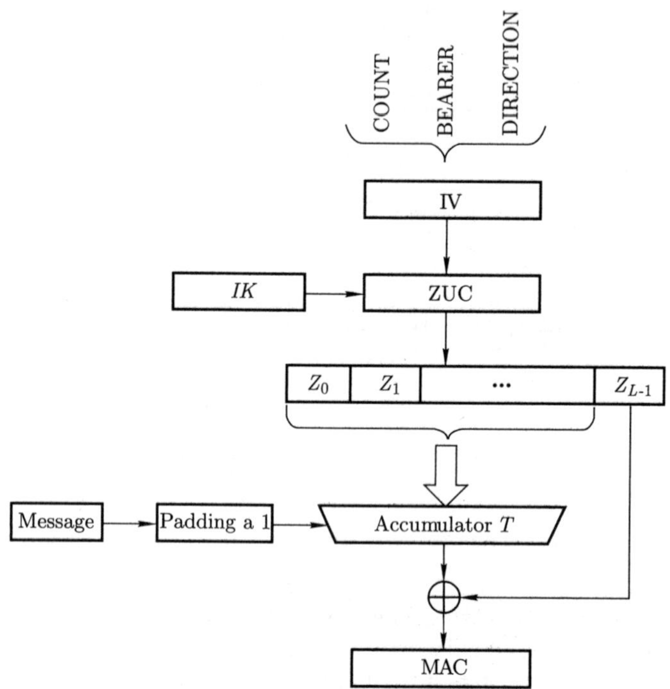

Fig. 2.14 The process of the EIA3-version1.5 MAC computation.

2.5.6 The security limitation of the authentication algorithm in LTE

It is noted that in LTE systems, the authentication code of any message is defined to be in 32 bits. This remains the same in EIA1, EIA2, and EIA3. Because with a birthday attack, a collision can be found with computation effort of 2^{16}, or a forgery authentication code can be found for any given message with computation effort of 2^{32}. This limitation however is an inherent problem in LTE and cannot be changed, unless the industry standard is to be changed.

2.6 Security issues in femtocell

With the increased demand on the network bandwidth of mobile communications, signal frequency becomes higher and higher, and the coverage of each base station also becomes smaller. This means that when being viewed as cellular networks, the cells become smaller. Moreover, for indoor mobile devices, good signal coverage needs more base stations to be installed. This also incurs high cost both of installation and of maintenance.

A new system called "femtocell" was proposed[25] to solve the problem of signal coverage limitations for wideband mobile networks. The femtocell introduces a kind of micro-station, called Home NodeB (HNB) as in 3G networks, or Home eNodeB (HeNB) as in LTE networks, is a good solution to compensate the limitation of indoor signal coverage. The network structure of femtocells can be depicted in Fig. 2.15.

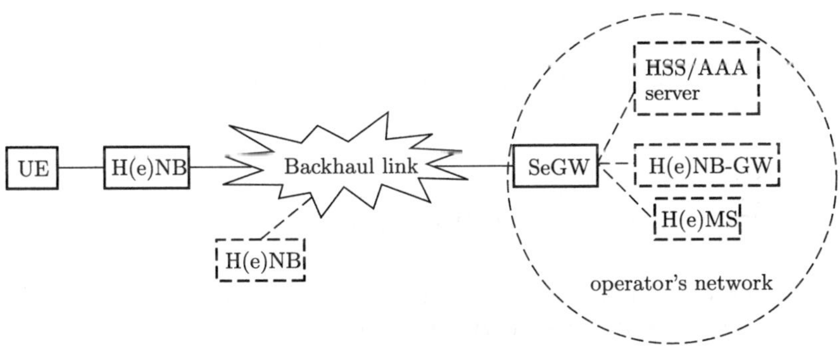

Fig. 2.15 The network structure of femtocell.

From Fig. 2.15 it is seen that the Home (e)NodeB (H(e)NB) is a micro station serving for mobile users (e.g. UE), where the H(e)NB is connected to an operator's network via the Internet (LAN or ADSL), which is not trusted by the network operator.

There are different sources of security threats to the femtocell system. First, the environment where the H(e)NB's are installed is not trusted by the operator, this is because the H(e)NB's are often installed in places out of control by the operator, such as private homes, private hotels, or buildings, where public access is limited, so that the access to the H(e)NB's by the operators becomes difficult. Second, the connection from the H(e)NB's to the operator's core networks is not controlled and hence trusted by the operator, because the connection is often through public networks such as the Internet. Third, the number of H(e)NB's can become quite large compared with the traditional base stations, which may induce unpredictable security threats, such as colluding attack from the H(e)NB's on the network (the public network or the operator's network).

When an H(e)NB is installed in an untrusted environment, users may install illegally manufactured H(e)NB's, or to manipulate some originally legal H(e)NB's for perhaps malicious purposes. The illegal modification/installation of H(e)NB's may abuse some victim mobile users when being serviced, or provide wrong information to the operator which will confuse the accounting process. Due to the H(e)NB's are connected to the operator's core network via public networks, some attacks to the public networks may be modified to attack the femtocell system, either the end H(e)NB's or the operator's core networks. The security threats due to a large number of H(e)NB's can come from the process of manufacture control, since H(e)NB's need to have a key (a shared symmetric key or a private key corresponding to a public key certificate) burned when they are being manufactured.

There are many standard requirements for the manufacturing and installation of H(e)NB's, but from the point of view of attackers, there is no point to follow those requirements. Technical requirements and rules do not provide good solutions to practical security threats.

Security threats should taken into consideration the situation when an H(e)NB is also integrated as a security gateway of a home-based wireless network, this is a specific scenario in the concept of Internet of Things (IoT), and is not discussed in depth here.

2.7 Privacy issues in cellular networks

With the increased number of users in cellular networks, and with the increased demand on services provided by the cellular and mobile networks, user privacy becomes a notable issue. Many kinds of information that were previous not treated as private now need to be classified as private. For example the information about the location of a mobile user is such an instance. When the number of users was small, it was not a big concern. But now the location information may be illegally used to trace a user and becomes a kind of private information and hence needs to be protected. The privacy issue is particularly important in some of the services such as mobile device assisted medical/therapy systems.

From cryptographic approach, there are techniques closely related to privacy issues. For example, blind signature, anonymous signature, and zero knowledge proof in some sense. However, those techniques need to be applied to practical application environment to solve practical privacy requirements. There seem to have some confliction in the privacy problem. A natural understanding on the privacy issue is the incapability to link an identity to some other information, e.g. medical record. However, when a particular medical record is needed, there must be a valid identity provided for the purpose of authentication. On one hand, how to use an identity information for the authentication purpose and on the other hand to protect the identity information against unauthorized access is a big challenge. Very often the problem becomes simpler if a trusted third party (TTP) is involved. The limitation is communication and computation bottleneck of the TTP. So far technical solutions on the privacy issues are far from being satisfactory in many practical applications, this is an area needing to be studied further and it has enormous practical applications.

2.8 Security issues of mobile devices

The security of cellular networks and communication systems tend to provide better security services, including the security protocols and security algorithms. However, there are security threats on the mobile equipments not caused by communication security protocols or the security algorithms. These kinds of security threats include the operating system malfunctioning and the loss of mobile devices.

The advanced mobile devices today are not just cellphones, they include laptop computers. Even considering only the cellphones, many of the cellphones have many functionalities same as in a computer, including an operating system and many applications. Most popular operating systems for smart cellphones include Symbian, Windows CE, Windows Mobile, and palm OS. Different operating systems may have different behavior in different aspects, for example with respect to the memory management, response time, and energy management. They all suffer security threats caused by mobile worms, viruses, and unauthorized access. Since the mobile operating system cannot have sophisticated antivirus software which would consume much of the operation resource, mobile operating systems are more fragile against security threats than many other operating systems for computers. Therefore, lightweight and efficient antivirus software for cellphones will be in a high demand.

Another security threats for mobile communication is the loss of mobile devices, in particular the missing, stolen, damage of cellphones. In this case the users will lose all the important information stored in the cellphones. To reduce the loss caused by mobile device loss, secure and timely data backup services are important. In the case of a cellphone being stolen, it may risk the privacy of the cellphone owner. The proper protection of the data stored in cellphones is important. Therefore, with respect to the data security and protection of mobile devices, on one hand the data stored in mobile devices needs to be securely protected, and on the other hand there should be a good way for secure data backup and recovery mechanism.

2.9 Concluding remarks

The technology of wireless and mobile communication evolves, and the concept of cellular networks also evolves. Today with the concept of ubiquitous computing becoming popular and the technology becoming mature, the concept of ubiquitous networks also emerges, which seems to cover the traditional concept as well as many advanced techniques of cellular networks. The new concept also means more services and inevitably more security threats and challenges as well.

This chapter tries to give a general picture about different aspects of security problems in cellular networks and communications. It was realized that there are so many techniques with respect to the information security problems in cellular networks and communications, and it ended up with a very brief introduction and has limited coverage. It was hoped that this introduction could help some readers to know the security challenges and motivate their interest to find good solutions.

References

[1] Farley T (2007) The Cell-Phone Revolution. American Heritage of Invention & Technology, 22(3): 8 – 19.

[2] Goldberg I, Wagner D, Green L (1999) The (Real-Time) Cryptanalysis of A5/2. Rump session of Crypto'99.

[3] Barkan E, Biham E, Keller N (2003) Instant Ciphertext-Only Cryptanalysis of GSM Encrypted Communication. Proceedings of Crypto 2003, LNCS 2729: 600 – 616. Springer-Verlag, Berlin.

[4] Ekdahl P, Johansson T (2003) Another attack on A5/1. IEEE Transactions on Information Theory, 49(1): 284 – 289.

[5] Barkan E, Biham E (2006) Conditional Estimators: An Effective Attack on A5/1. Selected Areas in Cryptography 2005, LNCS 3897: 1 – 19. Springer-Verlag, Berlin.

[6] Matsui M, Tokita T (2000) MISTY, KASUMI and Camellia Cipher Algorithm Development. Mitsibishi Electric Advance (Mitsibishi Electric corp.) 100: 2 – 8.

[7] Dunkelman O, Keller N, Shamir A (2010) A Practical-Time Related-Key Attack on the KASUMI Cryptosystem Used in GSM and 3G Telephony. CRYPTO 2010, LNCS 6223: 393 – 410, Springer-Verlag.

[8] Dunkelman O, Keller N, Shamir A (2010) A Practical-Time Attack on the A5/3 Cryptosystem Used in Third Generation GSM Telephony. Cryptology ePrint Archive: Report 2010/013.

[9] Specification of the 3GPP Confidentiality and Integrity Algorithms; Document 1: f_8 and f_9 specifications (3GPP TS35.201 Release 6). Available at http://www.3gpp.org/ftp/Specs/html-info/35201.htm. Accessed 10 November, 2011.

[10] 3GPP System Architecture Evolution (SAE); Security architecture (3GPP TS33.401 Release 9). Available at http://www.3gpp.org/ftp/Specs/html-info/33401.htm. Accessed 10 November, 2011.

[11] Specification of the 3GPP Confidentiality and Integrity Algorithms UEA2 & UIA2. Document 1: UEA2 and UIA2 specifications. Available at http://cryptome.org/uea2-uia2/uea2-uia2.htm. Accessed 10 November, 2011.

[12] Specification of the 3GPP Confidentiality and Integrity Algorithms; Document 2: KASUMI Specification (3GPP TS35.202). Available at http://www.3gpp.org/ftp/Specs/html-info/35202.htm. Accessed 10 November, 2011.

[13] Specification of the 3GPP Confidentiality and Integrity Algorithms UEA2 & UIA2. Document 2: SNOW 3G specification. Available at http://www.3gpp.org/ftp/Specs/html-info/35216.htm. Accessed 10 November, 2011.

[14] ETSI/SAGE Specification. Specification of the 3GPP Confidentiality and Integrity Algorithms EEA3 & EIA3. Document 1: EEA3 and EIA3 Specification; Version: 1.0; Date: 18th June, 2010. http://www.gsmworld.com/our-work/programmes-and-initiatives/fraud-and-security/gsm_security_algorithms.htm. Accessed 10 November, 2011.

[15] ETSI/SAGE Specification. Specification of the 3GPP Confidentiality and Integrity Algorithms EEA3 & EIA3. Document 2: ZUC Specification; Version: 1.0; Date: 18th June, 2010. http://www.gsmworld.com/our-work/programmes-and-initiatives/fraud-and-security/gsm_security_algorithms.htm. Accessed 10 November, 2011.

[16] ETSI/SAGE Specification. Specification of the MILENAGE-2G Algorithms: an Example Algorithm Set for the GSM Authentication and Key Generation Functions A3 and A8. Version 1.0. May, 2002. http://www.gsmworld.com/our-work/programmes-and-initiatives/fraud-and-security/gsm_security_algorithms.htm. Accessed 10 November, 2011.

[17] Dai Watanabe, Alex Biryukov, Christophe De Cannière (2004) A distinguishing attack of SNOW 2.0 with linear masking method. In Selected Areas in Cryptography 2003. LNCS 3006: 222–233, Springer-Verlag, Berlin.

[18] Nicolas Courtois and Willi Meier (2003) Algebraic Attacks on Stream Ciphers with Linear Feedback, In Advances in Cryptology-EUROCRYPT 2003, LNCS 2656: 346–359, Springer-Verlag, Berlin.

[19] Knudsen L R, Preneel B, and Rijmen V (2010) Evaluation of ZUC, ABT Crypto, Version 1.1, May, 2010.

[20] Cid C, Murphy S, Piper F, and Dodd M (2010) ZUC Algorithm Evaluation Report, Codes & Ciphers Ltd., 7 May, 2010.

[21] Hongjun Wu, et.al. (2010) Cryptanalysis of Stream Cipher ZUC in the 3GPP Confidentiality & Integrity Algorithms 128-EEA3 & 128-EIA3, presented at the Rump session of Asiacrypt 2010, Singapore.

[22] Fuhr T, Gilbert H, Reinhard J R, and Videau M (2010) A forgery attack on the candidate LTE integrity algorithm 128-EIA3 (updated version), Cryptology ePrint Archive, Report 2010/618, 2010, Available at http://eprint.iacr.org/. Accessed 10 November, 2011.

[23] ETSI/SAGE Specification. Specification of the 3GPP Confidentiality and Integrity Algorithms 128-EEA3 & 128-EIA3. Document 1: 128-EEA3 & 128-EIA3 Specification; Version: 1.5, 4th January, 2011. Available at http://www.gsmworld.com/our-work/programmes-and-initiatives/fraud-and-security/gsm_security_algorithms.htm. Accessed 10 November, 2011.

[24] ETSI/SAGE Specification. Specification of the 3GPP Confidentiality and Integrity Algorithms 128-EEA3 & 128-EIA3, Document 2: ZUC Specifica-tion, Version: 1.5, 4th January, 2011. Available at http://www.gsmworld.com/our-work/programmes-and-initiatives/fraud-and-security/gsm_security_algorithms.htm. Accessed 10 November, 2011.

[25] Femtocell forum, http://www.femtoforum.com/femto/.

Chapter 3
Security in Wireless Local Area Networks

Chao Yang[1] and Guofei Gu[2]

Abstract

Wireless Local Area Networks (WLAN) allow end users to wirelessly access Internet with great convenience at home, work, or in public places. WLANs are currently being widely deployed in our real life with great success. However, it is still in its infant stage as long as security is concerned. In this chapter, we briefly overview the security issues in the Wireless Local Area Networks (WLAN). After a short introduction to the background of WLAN, we present WLAN security requirements and categories of current real-world WLAN attacks. We then describe some details of several representative WLAN security protocols such as WEP, WPA, WPA2, and WAPI. We also survey security issues of the WLAN access points such as rogue access points and evil twin attacks. Finally, we overview other security mechanisms that can be used to enhance WLAN security, including Wireless Firewalls, Wireless VPN, and Wireless IDS.

3.1 Introduction to WLAN

3.1.1 WLAN Background

With people's huge demand of accessing the Internet wirelessly and the wide deployment of Wi-Fi equipments, wireless local area networks (WLANs) are nearly everywhere and are easy to find no matter at the coffee shops, restaurants, hotels, airports, private home, enterprises, universities, or government

1 Texas A&M University, College Station, Texas, USA. E-mail: yangchao@cse.tamu.edu.

2 Texas A&M University, College Station, Texas, USA. E-mail: guofei@cse.tamu.edu.

facilities. A wireless local area network is a network linking two or more devices by using wireless distribution methods (typically spread-spectrum or orthogonal frequency-division multiplexing radio), and usually providing a connection through an access point to the wider Internet[14]. In practice, a WLAN consists of two main categories of components: wireless-enable clients such as laptops, PDAs and smart phones equipped with wireless cards and wireless access points (APs) such as wireless routers. The main functions of the wireless access points are to receive and transmit radio frequencies for the wireless clients.

To achieve the goal of standardizing the implementations of WLANs, IEEE LAN/MAN Standards Committee (LMSC) creates and maintains a set of IEEE 802.11 standards[6] for WLANs. The services specified in IEEE 802.11 for the implementations of WLANs include both radio standards and networking protocol standards. These standards guarantee the acceptability of the wireless connectivity to fixed stations, portable stations, and moving stations within the specific area of the network.

3.1.2 WLAN Architecture

In an 802.11 WLAN, all the components belonging to the WLAN are referred to as "stations". A set of stations can form a basic building block called "basic service set" (BSS). The stations in the basic service set communicate with each other obeying the same networking protocol under the same, shared wireless medium, which may generate medium access collisions. Every BSS has a unique identification (ID) called BSSID, which is the MAC address of the access point servicing the BSS. Multiple BSSs connected through a wired or wireless distribution system can form an extended service set (ESS). Each ESS also has an ID called service set identifier (SSID) which can be up to 256 characters long now.

From the viewpoint of the network architecture, WLANs can be divided into two categories: infrastructure-based WLANs and Ad Hoc WLANs. The majority of current WLANs are infrastructure-based, such as IEEE 802.11 WLANs. In an infrastructure-based WLAN, each device connects to the network by establishing a wireless connection to a pre-installed base station to transmit and receive packets. The base stations in the WLAN are usually connected through high bandwidth wired connections. In this way, the communication typically takes place between the wireless clients and the base station rather than directly between the wireless clients. The main aim of the infrastructure-based networks is to provide wireless services to users in a fixed network area. An example of an infrastructure-based WLAN can be found in Fig. 3.1.

Unlike the infrastructure-based wireless network, the stations in an ad hoc network communicate with each other directly peer to peer (P2P) without the need of any pre-existing fixed infrastructure or base stations. In this way,

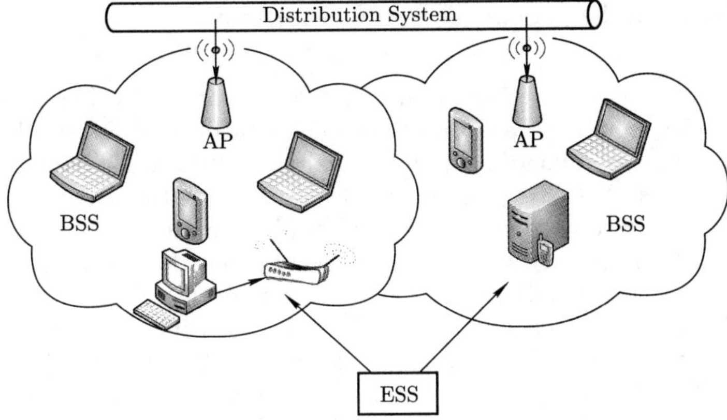

Fig. 3.1 An example of an infrastructure-based WLAN.

the Ad Hoc network can offer the service to users without the constraints of certain geographical situations. An example of an Ad Hoc WLAN can be seen in Fig. 3.2.

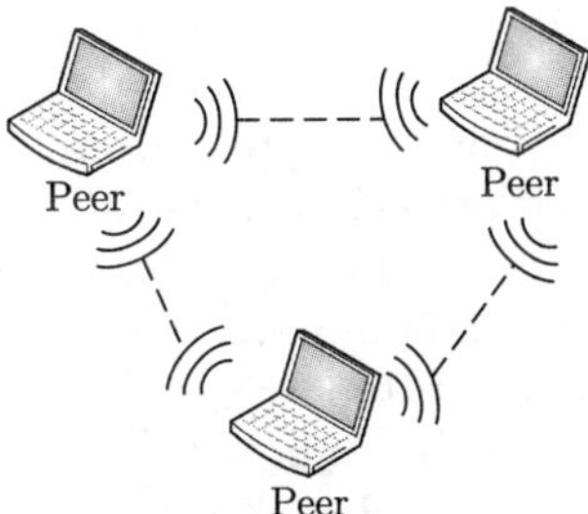

Fig. 3.2 An example of an Ad Hoc WLAN.

3.1.3 WLAN Applications

Current applications of WLANs have been extended into many areas such as LAN extension, public service, multimedia transmission, and mobile communication. WLANs are broadly being utilized from personal home networks to public places such as airline lounges, coffee shops, restaurants, stores and libraries, also ranging from personal service such as mobile IP and VoIP to public business such as education, healthcare, hospitality, financial industries, and public safety.

3.2 Current State of WLAN Security

The ubiquity, convenience and powerful strength of WLANs are not merely enticing to legitimate users but also to malicious attackers. Especially, attackers can utilize the vulnerabilities in the existing authorization and authentication policies in WLANs and the broadcast nature of the wireless communication to greatly compromise the security of legitimate wireless users. Thus, wireless LAN security has become a serious concern for an increasing number of wireless organizations. According to reference [44], nearly two-thirds (61%) of people consider security as the second most important WLAN characteristic after reliability (64%) and nearly half (49%) describe the ability to simplify WLAN security deployment as "very important". In this section, we will show a brief outlook of the current state of WLAN security.

3.2.1 WLAN Security Requirements

WLAN security is an important, dynamic, and even evolving topic. Novel threats, attacks, technologies and solutions are emerging almost every day. However, although diverse WLANs may have different infrastructure components and support distinct practical applications, to be effective, stable, and trustworthy, the security requirements of the WLANs essentially fall into the following five broad categories: confidentiality, authentication, access control, integrity, and intrusion detection and prevention.

- *Confidentiality*: Confidentiality prevents the disclosure of the data or information to unauthorized individuals or systems, when that information is transmitted across the shared communication medium. Confidentiality can be achieved through the utilization of encryption techniques to encode the information in a manner so that the information can only be decoded, understood and analyzed by the authorized parties.
- *Authentication*: Authentication provides a service that verifies and confirms the authenticity of a sender or receiver's identity that it claims to be. Essentially, robust authentication mechanism in the WLANs not only ensures that the information can be transmitted from/to the authentic entities in the two-side parities of the communication, but also avoids these information to be interfered or impersonated by a third party. Without such an authentication mechanism, attackers can gain full access to the information transmitted in the WLANs or even control the WLANs.
- *Access Control*: Access control service enables an authority to grant authorized users the corresponding access right to the resources in the WLANs. In this way, sophisticated implementations of access control policies in the WLANs allow for granting different users or groups with different security settings and with different levels of access rights to the resources after authenticating these users' or groups' identities.

- *Integrity*: Integrity assures the consistency of the data when it is transmitted in the WLANs. This requirement is also usually achieved by the utilization of encryption techniques. Strong integrity is essentially crucial for wireless traffic, as wireless network packets can be easily intercepted, modified, or even compromised by the attackers in the WLANs due to the broadcast nature of the wireless communication.
- *Intrusion Detection and Prevention*: Due to the continually increasing attacks to the WLANs, in addition to the above requirements, a robust WLAN also needs to provide wireless intrusion detection and prevention services (Wireless IDS/IPS). These services can identify and remove threats, but still allow neighboring WLANs to co-exist while preventing clients from accessing each other's resources[24]. It involves detecting rogue access points, regulating network access and defending against wireless Denial-of-Service (DoS) attacks.

To effectively and efficiently meet the above security requirements in the WLANs, it is significant and indispensable to design and implement robust security policies for the WLANs. These policies should not only layout the security schemas for the installations, managements, and usage procedures, but also be flexible in terms of the supported technologies and functions. Whenever the security policies are implemented in terms of these security challenges in the WLANs, deeply understanding WLAN specific vulnerabilities and existing attacks will be necessary and beneficial to designing more robust security policies.

3.2.2 Real-World WLAN Attacks

As mentioned before, despite the productivity and convenience that the WLAN offers, the improper human configurations or the operations, and the vulnerabilities in the existing WLAN security policies can still be utilized by the attackers to make legitimate wireless users at a risk. "To advance irresistibly, push through their gaps." In order to design more robust security mechanism and more powerful defense methods to enhance the WLAN security, it is very useful and meaningful to understand current real-world WLAN attacks. Although attacks against WLAN technologies are increasing in number and sophistication over time, we can summarize most current real-world WLAN attacks into the following categories: deauthentication, eavesdropping and interception of wireless traffic, traffic jamming, brute force attacks against access point passwords, attacks against security protocols and misconfiguration.

- *Deauthentication*: This kind of attacks attempt to defeat the authorization mechanism in WLANs. By launching this kind of attacks, the attackers can steal legitimate wireless users' identities or authorized wireless access points' deployment rights to mimic as authenticated users or deploy rogue access points without going through security process and review.

- *MAC Spoofing*: By modifying the wireless client's MAC address, the attackers can bypass the MAC filtering policies widely utilized in the most current wireless systems. Specifically, many wireless systems can use a white list of MAC addresses to authorize the wireless clients. Only the wireless clients whose MAC addresses are in the white list can gain access to the network. However, by utilizing some software that can make a wireless client to pretend to have any customized MAC address[18], the attacker can easily get around that hurdle.
- *IP Spoofing*: By modifying the source IP address contained in the packet header, an attacker can evade IP address based authentication and pretend itself to be a legitimately authenticated user who is communicating with others.
- *Rogue Access Points*: Rogue access points are unauthorized access points that are deployed in the WLANs. In this way, the unauthorized clients can gain the open access to the WLAN through the rogue access points. Also, these rogue access points can also be settled as "honeypot" or "phishing" access points to achieve attackers' malicious goals.

- *Eavesdropping and Interception of Wireless Traffic*: This kind of attacks can eavesdrop or intercept legitimate wireless traffic by compromising the legitimate users' wireless communication channel. Through this kind of attacks, the attackers could achieve all the sensitive and important information sent by the legitimate users.
  - *Traffic Eavesdropping*: Attackers can break the confidentiality of the data by eavesdropping the whole WLAN. Due to the broadcasting nature, all the information is passing from the network interface cards (NIC) across a communication medium and the centralized device intentionally radiates the network traffic into space. In this way, an attacker can simply utilize some wireless network sniffers such as Kismet[7], Wellenreiter[11], Airtraf[3] and Airfart[1], to eavesdrop the wireless traffic in the whole WLAN. In the WLANs, traffic eavesdropping is typically the first step for an attacker to launch other attacks.
  - *Man-in-the-middle Attacks*: In this attack, an attacker can sit in the middle of the two-way communicating parties. In this way, by successfully cheating the senders and receivers that they are communicating under a private and reliable connection channel, the attacker could not only obtain all the transmitted information, but also intercept, modify and even impersonate the communication. Especially, evil twin attack is one of the representative man-in-the-middle attacks. It is a term for a rogue Wi-Fi access point that appears to be a legitimate one offered on the premises, but actually has been set up by a hacker to eavesdrop on wireless communications among Internet surfers[5].
  - *Network Injection*: In this kind of attacks, an attacker can inject bogus network traffic into the legitimate traffic. By inserting this bogus traffic, the attacker could achieve malicious goals like sending

re-configuration commands to the access points to fully control them.

- *Session Hijacking*: This kind of attacks can be achieved by stealing a legitimate authenticated conversation session ID. As a result, the attacker could control the whole conversation session when it is still going on.

- *Traffic Jamming*: The goal of this kind of attacks is to heavily consume the bandwidth of the WLAN in order to overwhelm legitimate traffic. This kind of attacks can be achieved by flooding either valid or invalid messages, or high radio frequency signals.
  - *Denial of Service (DoS) Attacks*: Denial of service attacks are also easily applied to wireless networks, where legitimate traffic cannot reach the destinations due to the flooding of high-frequency radio signals or messages. Since the high bit rates of WLANs can overwhelm low bit rates of WLANs, an attacker can easily launch a denial of service attack by using a proper equipment that can flood higher radio frequency signals, corrupting all other legitimate signals until the whole WLAN ceases to function. In addition, an attacker can also use a wireless device to flood other wireless clients with bogus packets to create a denial of service attack.
  - *Spam Attacks*: Like spam in the traditional Internet security that can consume bandwidth and generate phishing attacks, attackers can also launch spam attacks by flooding spam messages over the whole wireless network channels. In this way, legitimate users cannot obtain normal service afforded by the WLAN due to the overflowing spam messages.
- *Brute Force Attacks Against Access Point Passwords*: Since most access points only use a single shared password with all connecting wireless clients, attackers can use brute force dictionary attacks to compromise this password by testing every possible password. As a result, the attacker could control the access point and even take over the whole WLAN.
- *Attacks Against Security Protocols*: To meet the security requirements, 802.11 standards have designed and utilized different security protocols such as Wired Equivalent Privacy (WEP) and Wi-Fi Protected Access (WPA). However, the vulnerabilities of these standards have been utilized by the attackers to crack them. For example, there are several WEP crackers such as AirSnort[2], Wepcrack[12] and Wep tools[13], which can be used by attackers to compromise WEP protocol.
- *Misconfiguration*: Many WLAN attacks are generated due to the limited security knowledge of the administrators of the WLANs, and human misconfigurations or improper operations to the access points. For example, access points are usually sold with an unsecured and common configuration with a goal of easing consumers' usages. Unless administrators with certain wireless security knowledge and properly configure the access points, these access points will remain at a high risk for being attacked. However, many studies (e.g., reference [26]) have pointed out that

many users would keep the default security configurations of the access points when they are deploying their WLANs. Obviously, these WLANs are very vulnerable and can be easily compromised by attackers.

3.3 WLAN Communication Security

After knowing about a bunch of real-world WLAN attacks, we also need to understand the advantages and weakness of existing WLAN security standards that are deployed to satisfy WLAN security requirements. Thus, this section will describe the security details of two existing representative IEEE 802.11 security standards — Wired Equivalent Privacy (WEP) and Wi-Fi Protected Access (WPA). Also, this section will give a brief introduction to other standards such as 802.1x, 802.11i (WPA2), and WAPI.

3.3.1 WEP Protocol

Wired Equivalent Privacy (WEP) is the first IEEE 802.11 security protocol, which was designed in September 1999. The main goal of this protocol is to guarantee the confidentiality, authentication and integrity by implementing encryption techniques in the MAC layer to protect link-level data communication security between the clients and the access points. Basically, WEP is implemented from the initial connection between the clients and the APs. The clients can only successfully connect to the APs by using the correct passwords. Also, WEP achieves the security goals by encrypting the transmission so that only the receivers who own the correct decryption key can decrypt the transmitted information.

3.3.1.1 WEP Framework

WEP utilizes RC4 encryption algorithm, CRC-32 (Cyclic Redundancy Code) checksum algorithm, and a pre-established shared secret key (the base key) to encrypt the transmission between the clients and APs. The original base key with a fixed value was 40 bits long. The key had been increased by most manufactures to 104 bits with a security concern.

Furthermore, WEP utilizes a generated traffic key which is the base key added with an initialization vector (IV). The initialization vector is a randomly-generated 24-bit sequence, converting the original 104-bit key to a new 128-bit key. In this way, since the values of the IV vary when different packets are generated, the encryption keys are also different for encrypting different packets. Thus, the same plaintext may generate different cipher text at different times.

3.3.1.2 WEP Vulnerabilities

Nowadays, WEP is no longer considered as a secure mechanism for WLAN, because it contains several vulnerabilities and can be compromised by the attackers. The major WEP vulnerabilities can be summarized into the following four categories[41]:

- *No forgery protection*: There is no forgery protection provided by WEP. Even without knowing the encryption key, an adversary can change 802.11 packets in arbitrary, undetectable ways, deliver data to unauthorized parties, and masquerade as an authorized user. Even worse, an adversary can also learn more about the encryption key with forgery attacks.
- *No protection against replays*: WEP does not offer any protection against replays. An adversary can create forgeries without changing any data in an existing packet, simply by recording WEP packets and then retransmitting later. Replay, a special type of forgery attack, can be used to derive information about the encryption key and the data it protects.
- *Misusing the RC4 encryption algorithm*: Although RC4 encryption algorithm should not be blamed, WEP misuses the RC4 encryption algorithm in such a way to expose the protocol to the weak key attacks. An attacker can utilize the WEP IV to identify RC4 weak keys, and then use known plaintext from each packet to recover the encryption key.
- *Reusing initialization vectors*: It is known that if the same traffic key should not be used twice for a stream cipher such as RC4. Since the length of the IV in IEEE 802.11 WEP is 24, there are only 16 777 216 possible values of the IV. In a large and busy network, an access point may exhaust the space of IVs and thus reuse the same IV after several hours. Furthermore, due to the well-known birthday paradox, for a 24-bit IV, there is a 50% probability the same IV will repeat after 212 (4096) packets. Thus, WEP enables an attacker to decrypt the encrypted data without ever learning the encryption key or even resorting to high-tech techniques by using the brute force attack.

3.3.1.3 WEP Attacks

Due to the above vulnerabilities in WEP, attackers have already launched attacks on WEP by compromising these vulnerabilities. This section describes the following three major attacks on WEP: Brute force attack, Key Stream Re-uses, and Weak IV attacks[23].

- *Brute force attack*: As mentioned before, there are around 17 million possible values of the IV, the brute force attack will try all possible keys either by manually or by the computers until the correct one is found. Attackers can utilize the computers to find the key within the time period of less than several days by a continuous search.
- *Key Stream Re-use attacks*: According to the policy of the Shared Key Authentication in WEP, the authenticator will first send a clear text to the supplicant also known as authentication peer. Then, the supplicant will be authenticated by replying with the correctly encrypted message

of the text. If an attacker can steal the ciphertext and plaintext pair by snooping the authentication communication, the attacker can simply recover the key stream by using RC4 algorithm on the ciphertext and plaintext pair. Once the attacker successfully recovers the key stream, he can decrypt all the data which is associated with that key stream.

- *Weak IV attacks*: By collecting sufficient data packets using weak IVs, the attacker can re-calculate the accurate WEP key[27]. Specifically, a single weak IV reveals a correct key byte 5% of the time. By gathering a high number of statistics (IVs), the most probable key may be calculated within several days.

3.3.1.4 WEP Cracking Tools

Due to WEP's vulnerabilities, many public tools have been developed to crack WEP. This section will briefly introduce several WEP cracking tools[40].

- *AirSnort*: One of the most famous WEP cracking tools is AirSnort[2]. By displaying an intuitive human-machine interface, AirSnort is very convenient for people to use to discover networks and crack WEP. Besides cracking WEP, AirSnort can also be used to dump wireless packets and to save them as pcap-format files.
- *Wepcrack*: As one of the first few WEP cracking tools implementing theoretical attacks into practice, Wepcrack[12] consists of a collection of Perl scripts such as WEPcrack.pl, WeakIVGen.pl, and prism-getIV.pl. It can collect packets with initialization vectors (IVs) and save the weak IVs in a log file called IVFile.log. Then, attackers can simply use the following command to crack WEP protocol: (assuming the wireless network interface is wlan0)

```
root:# tcpdump - i wlan0 -w - | perl prism-getIV.pl
```

- *Wep_tools*: Wep_tools[13] is a WEP cracking toolkit implementing brute-force and dictionary attacks. By compromising the 40-bit WEP-from-passphrase generation algorithm, it is efficient to crack original 40-bit WEP keys. For the 128-bit WEP keys, attackers are limited to launch dictionary attack by using practical terms. Wep_tools can be run on Linux machines using the following command[40]:

```
root:# ./wep_crack
Usage: ./wep_crack [-b] [-s] [-k num] packfile [wordfile]
-b       Brute force the key generator
-s       Crack strong keys
-k num   Crack only one of the subkeys without using a key generator
```

3.3.2 WPA Protocol

As an enhanced WLAN security protocol, Wi-Fi Protected Access (WPA) is invented by Wi-Fi Alliance (WFA) in the year of 2002 to improve the initial security standard WEP. Essentially, WPA is implemented by designing more complex encryption and authentication methods in place of merely using WEP's basic RC4 encryption.

WPA contains two modes: Enterprise/commercial WPA and Personal/ WPA-PSK (Pre-Shared Key) WPA. In Enterprise mode, WPA functions as a Remote Authentication Dial In User Service (RADIUS) server. It provides centralized Authentication, Authorization, and Accounting (AAA) management for computers to connect and use a network service. In Personal mode, it utilized Pre-Shared Key (PSK) containing the network SSID and the WPA key generated by the access point to provide authenticity to wireless networks.

3.3.2.1 WPA Framework

WPA achieves the goal of designing a more secure wireless standard by mainly using the Temporal Key Integrity Protocol (TKIP) and Message Integrity Check (MIC). In the TKIP protocol, it has two different keys: a 128-bit key, which is used by a mixing function to produce a per-packet encryption key, and a 64-bit key, which is used to guarantee message integrity.

As discussed before on WEP's vulnerabilities, one major weaknesses of WEP was the small size of its initialization vector. In TKIP, the size of IV is increased from 24 to 40, which can effectively reduce the probability of generating key collisions. In addition, every key in the TKIP has its own fixed lifetime. The key will automatically be replaced when the key reaches its lifetime. Although WPA also uses RC4 algorithm like WEP, the per-packet key mixing function and re-keying mechanism in the WPA can guarantee that keys are frequently updated when using RC4. With the larger key size and the dynamic key encryption method, WPA can defend against stronger attacks. Also, instead of using Cyclic Redundancy Check (CRC) in the WEP standard, WPA guarantees the message integrity by using Message Integrity Check (MIC). The purpose of MIC is to prevent an attacker from capturing, altering and/or re-sending data packets. Essentially, it achieves this by appending 64 bits Cryptographic Message Integrity Code with the IV.

3.3.2.2 WPA Vulnerabilities

In general, WPA is a stronger encryption standard than WEP by using the TKIP protocol. However, it may still be an interim solution due to its several vulnerabilities, which will be described in this section.

Since WPA still utilizes the RC4 cipher stream algorithm, an attacker can also brute force two distinct RC4 keys to recover the 128-bit temporal key in WPA, known as temporal key recovery attack[34]. Once an attacker achieves the key, he can nearly do anything before current temporal key expires.

In Personal WPA mode, it utilizes the Pre-Shared Keys (PSKs) for the authentication rather than using a dedicated authentication server. Due to the broadcasting nature of the wireless device to create and verify a session key, the attacker could steal the information about the key by passively sniffing the wireless communication channel. Also, the attacker can launch an offline dictionary attack on the keys, when WPA tools are using handshake process for exchanging the data encryption keys between the access point and the end user. Thus, PSK, requiring simple deployments, is designed to meet the security requirement in the small and less critical wireless networks. However, the risk of using PSK can still not be neglected.

3.3.2.3 WPA Attacks

Similar to the situation of WEP, attackers also utilize WPA's vulnerabilities to launch their attacks. As the problem of the PSKs mentioned in the previous section, any key generated from a passphrase of less than about 20 characters is highly vulnerable to the offline PSK dictionary attack[35].

In addition, although WPA utilizes more sophisticated methods and protocols to prevent key attacks, the attacker can still launch an improved version of ChopChop attack[22] to decrypt the wireless traffic by sending customized packets to the network. In addition, WPA may also suffer from DoS attack. For example, when the WPA wireless device receives two packets of unauthorized data within one second from the same user, it will assume it is under attack and automatically shut down itself. In this way, the attacker can launch the DoS attack by rapidly repeating sending authentication packets to the wireless device.

3.3.3 Other Security Protocols

In addition to the above two traditional and representative WLAN security protocols, we also briefly introduce other security standards such as 802.1x, 802.11i (WPA2), and WAPI in this section.

3.3.3.1 802.1x

As part of the 802.11i standard, IEEE 802.1x protocol is designed for the Port-based Network Access Control (PNAC). It provides an authentication mechanism for the wireless devices to connect to a LAN or WLAN. It also guarantees the security requirement of the data transmission for the components that are connected with each other through different 802.11 LANs.

The 802.1x authentication system has three major components: a supplicant, an authenticator, and an authentication server. The supplicant is a wireless client device wishing to connect to the WLAN. The supplicant refers to the software running on the client that provides credentials to the authenticator. The authenticator is usually a network device (e.g., a wireless access point) that transmits this information between the supplicant and the authentication server. The authentication server is typically a network device, such as an Ethernet switch or wireless access point, running software to support the RADIUS and Extensible Authentication Protocol (EAP), which is defined in the 802.1x standard. In this way, the authenticator, validating and authoring the supplicant's identity, acts like a security guard to protect the WLAN.

3.3.3.2 802.11i (WPA2)

After the 802.1x standard, IEEE 802.11i, also known as WPA2, is an additional specification that is finalized in fall 2004 in order to provide replacement technology for WEP security in the WLAN. Generally, to provide enhanced WLANs' security, WPA2 defines data confidentiality, mutual authentication, and key management protocols.

Compared with WEP and WPA, one of the significant improvements of WPA2 is that it utilizes a single component, named as counter mode with CBC-MAC Protocol (CCMP), for authentication, key management and message integrity. CCMP is built based on an enhanced version of encryption algorithm — Advanced Encryption Security (AES), which is one of the most secured encryption standards. Specifically, CCMP consists of two components: Counter mode, used in AES to encrypt the data that provides data protection from unauthorized access, and Cipher Block Chaining Message Authentication Code (CBC-MAC) mode, creating a Message Integrity Check (MIC) code to provide message integrity. In addition, WPA2 use 802.1x or pre-shared keys (PSKs) to authenticate the wireless client and the authentication server. It also defines the Robust Security Network Association (RSNA) protocol to provide mutual authentications.

In brief, the comparison of WEP, WPA, and WPA2 can be summarized in Table 3.1.[36]

Table 3.1 The comparison of WLAN security protocols

Security Protocol	WEP	WPA	WPA2
Major Component	IV	TKIP	CCMP
Stream Cipher	RC4	RC4	AES
Key Size	40 bit	128 bit (encryption) and 64 bit (authentication)	128 bit
IV Size	24 bit	48 bit	48 bit
Key Management	Not Available	IEEE 802.1x/EAP	IEEE 802.1x/EAP/CCMP
Date Integrity	CRC-32	MIC	CBC-MAC

As shown in Table 3.1, the main advantages of the WPA2 standard can be listed as follows[30]:

- Providing more excellent security by using advanced encryption algorithms;
- Using stronger key management policies;
- Protecting against the man-in-the-middle attacks by using the two-way authentication process;
- Providing improved message integrity by using CBC-MAC.

Although WPA2 is designed to cover up for the weaknesses of WEP, it still has its own drawbacks. First, WPA2 is costly. Due to the requirements of the implementation of the advanced properties designed in WPA2 (e.g., CCMP), a lot of money and effort will be costed on upgrading existing hardware and software. Also, due to the need of bidirectional authentication between users and access points, WPA2 requires more hardware to achieve the security goal. Second, WPA2 is still vulnerable to DoS attacks[36]. Attackers can send large amount of authentication requests to the authentication server simultaneously so that the 8-bit space of EAP packet will be exhausted, leading the network under DoS attacks. Third, WPA2 is also prone to attacks such as security level rollback attack, reflection attack, and Time Memory Trade Off (TMTO) attack. Specifically, when Pre-RSNA and RSNA algorithms are both used in a single WLAN, an adversary can launch a security level rollback attack, avoiding authentication and disclosing the default keys[29]. Also, if a device is implemented to play the roles of authenticator and supplicant (in ad hoc networks, typically not in infrastructure networks), attackers can launch the reflection attack during the 4-Way Handshake. Current studies[33] also show that attackers can launch TMTO pre-computation attack, if they have sufficient knowledge about the WLAN so that they can successfully obtain the initial counter value used in the AES of CCMP.

3.3.3.3 WAPI

Besides internationally well-acknowledged WLAN security standards, to adapt to the rapid developments of Chinese WLANs and to meet the security requirements of Chinese wireless users, China has also finalized its own national WLAN security standard in 2003 — WLAN Authentication and Privacy Infrastructure (WAPI)[15]. According to WAPI protocol specification[16,17], WAPI consists of two modules: Wireless Authentication Infrastructure (WAI) and Wireless Privacy Infrastructure (WPI). Specifically, WAI is designed for the authentication process and key management and WPI is implemented to provide the data protection and integration service.

As the major module of WAPI, WAI[9] adopts port-based authentication architecture to authorize the credentials similar to 802.1x standard, including three components: the Authentication Supplicant Entity (ASUE), the Authentication Entity (AE), and the Authentication Service Entity (ASE). The process of the certificate authentication and key management in the WAPI can be illustrated in Fig. 3.3.

In the process of certificate authentication, AE first sends authentication activation packets to ASUE to active the entire authentication process. Once receiving the authentication activation from AE, ASUE verifies whether the activation packets meet ASUE's requirements. If so, ASUE will send an authentication request with its own certificate and an access request time to AE. Then, AE signs its own name on the ASUE's certificate, ASUE's access request time and its own certificate, and sends this information as the certificate authentication request to ASE. After the certificate request is successfully authenticated by ASE, AE will receive the certificate authentication

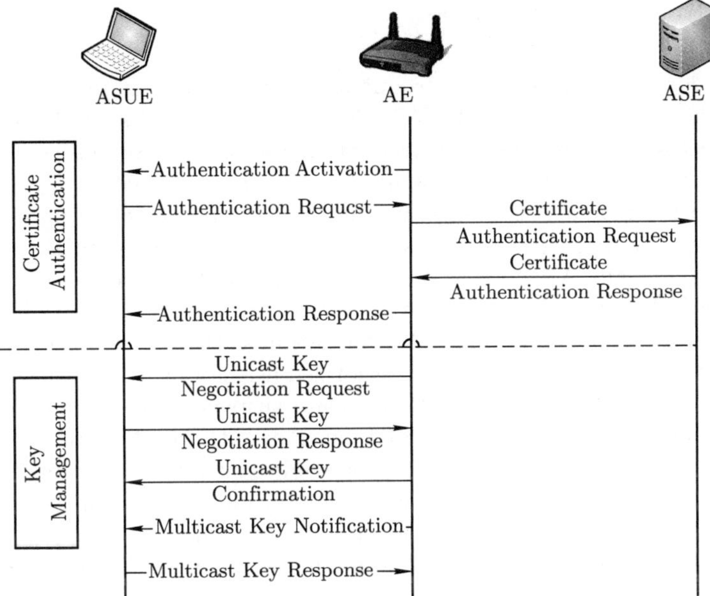

Fig. 3.3 Illustration of WAPI authentication process.

response from ASE and send it to ASUE. Finally, ASUE decides whether to access AE by checking the authentication response from ASE. From this Security in Wireless Local Area Networks architecture, we can see that WAPI supports the mutual authentication between ASUE (wireless clients) and AE (Access points).

In the process of key management, AE will first send a unicast key negotiation request including cryptography algorithms negotiation to ASUE. Once AE receives the agreement response of the negotiation request from ASUE, AE will send a unicast key confirmation to ASUE. After successfully building the agreement on the execution of the unicast key, AE will start the multicast key process, which utilizes the unicast session key for the encryption.

In short, as the first WLAN standard developed and owned by China itself, WAPI undoubtedly plays a very important role in the developments of the field of WLAN security in China.

3.4 WLAN Access Point Security

As one essential component in WLAN, access points, directly communicating with the end-users, need to be carefully deployed and protected. Thus, in this section, we mainly talk about security issues in the WLAN access points.

3.4.1 Rogue Access Points

As mentioned in Section 3.2.2, rogue access points are unauthorized access points that are deployed in the WLANs. The main purpose of deploying the rogue access points for the attackers is to get access of other users' resources. Specifically, the unauthorized clients can gain the open access to the WLAN through the rogue access points. In addition, the rogue access points can be utilized as honeypot access points to steal other users' credentials.

Existing rogue AP detection solutions can be mainly classified into two categories. The first category of the approaches monitors Radio Frequency (RF) airwaves and/or additional information gathered at routers/switches and then compares with a known authorized list. For example, AirDefense[19] scans RF from the Intranet APs to locate suspicious ones, and then compares specific "fingerprints" of the RF with an authorized list to verify. More specifically, for the scanning part, some studies such as[8,4,10] rely on sensors instead of sniffers to scan the RF; some studies like[20] propose a method to turn existing desktop computers into wireless sniffers to improve the efficiency. For the verification part, these studies verify MAC addresses, SSID, and/or location information of the AP by using an authorized list. However, these studies still have the risk of falsely claiming a normal neighbor AP as a rogue AP with a high probability. To solve this problem, they need to further verify whether such a rogue AP is indeed in the internal network.

The second category of approaches detects rogue access points by differentiating whether the clients come from wireless networks or wired networks. Essentially, if a client comes from a wireless network while it is not authorized to use wireless (comparing with an authorized list), the AP attached to this host is considered as a rogue AP. Some work such as[21,33,37,42,43] use statistical features (e.g., entropy, median, mean) on the traffic time (e.g., RTT) to distinguish the type of network. It is also possible to use the frequent rate adaptation in the wireless network to distinguish it with wired networks[25]. However, this line of work should solve the problem of falsely claiming an authorized wireless user who connects to Intranet with wireless networks. Thus, they may still need to further verify a wireless device is an authentic AP or not with some "fingerprints" from the authorized lists. To solve this problem, two hybrid studies[32,39] provide the technique to compare the fingerprints in the integrated systems.

3.4.2 Evil Twin Access Point

As one special type of rogue AP, an evil twin AP is essentially a phishing Wi-Fi access point (AP) that pretends to be a legitimate one (with the same SSID name). It is set up by an adversary, who can eavesdrop or modify wireless communications of users' Internet access. In the next paragraph, we briefly introduce three representative works that are aiming to detect evil twin attacks.

In reference [31], Jana and Kasera utilize the fact that different APs usually have different clock skews to detect unauthorized wireless access points. This work utilizes the fingerprint technique, which still needs a white list of the authorized access points. In reference[28], Han et al. utilizes time interval information to detect rogue APs. Specifically, it calculates the round trip time between the user and the DNS server to independently determine whether an AP is legitimate or not without the assistance from the WLAN operators. Song et al.[38] proposes a user-side evil twin detection technique by differentiating one-hop and two-hop wireless channels from the user side. This work exploits fundamental communication structures and properties of evil twin attacks in wireless networks and designs active, statistical, and anomaly detection algorithms to identify evil twin APs.

3.5 Other WLAN Security Issues

Besides the security standards such as WEP, WPA, 802.1x, 802.11i and WAPI that have been discussed previously, other security mechanisms such as Wireless Firewalls, Wireless Virtual Private Network (VPN) and Wireless Intrusion Detection System (IDS) can also be utilized to enhance the security of WLANs.

- *Wireless Firewalls*: Like traditional network firewalls, a wireless firewall functions as a barrier between the private network and the Internet to prevent external attacks to the internal network. A wireless firewall can protect an internal host or server from insecure Internet traffic by filtering out suspicious packets.
- *Wireless VPN*: A virtual private network (VPN) utilizes a public telecommunication infrastructure, such as Internet, to provide secured remote communication for the users to their private organization network. Since WLAN uses unlicensed frequency bands and can be easily accessible to outsiders either accidentally or with malicious intent, wireless networking provides an important area for VPN deployment and maintenance[40].

Compared with the physical restriction on the deployments of wired VPNs, wireless VPNs can be applicable and deployed to any WLAN, as long as a high level of security is concerned. Although the standard of 802.11i can guarantee the same security requirements as the wireless VPNs, the vulnerabilities in the implementations of the 802.11i standard could still make it less trustworthy. Thus, in an environment requiring a high level of security, besides traditional protocol standards such as WEP, WPA and WPA2, wireless VPNs, based on the Internet Protocol Security (IPsec) protocol, can still function as another safeguard to protect the security in the WLAN. In addition, in the case of point-to-point wireless links it is easier and more economical to deploy a network-to-network VPN than 802.11i-based defenses, including the RADIUS server and user credentials database, while using 802.11i with PSK and no 802.11x is not a good security solution for a high throughput network-to-network link[40].

- *Wireless IDS*: An intrusion detection system (IDS) is a device or software attempting to perform network intrusion detection and stop possible incidents/attacks by gathering and analyzing data. To protect WLAN security, IDSs have already been developed for the use on the WLAN, known as wireless IDSs. Similar to traditional IDSs, these wireless IDSs can recognize patterns of known attacks, identify abnormal network activity, and detect policy violations for WLANs by monitoring and analyzing network, user, and system activities. Also, like traditional signature based IDSs and anomaly-based IDSs, wireless IDSs can generate intrusion alters according to either the predefined signatures or the observed abnormal network behavior.

Wireless IDSs can be divided into centralized IDSs and decentralized IDSs. In a centralized wireless IDS, the central management system will combine and analyze all wireless data from each distributed individual sensor. In a decentralized wireless IDS, there are more than one device that both collect data and generate the intrusion alerts by analyzing the data.

3.6 Conclusion

In this chapter, we have discussed security issues and techniques in the Wireless Local Area Networks (WLAN). Essentially, we present a brief introduction of the WLAN background and the current state of WLAN security. Then, we provide some details on wireless security protocols and access point security. Finally, we also talk about other security mechanisms that can be used to enhance WLAN security including Wireless Firewalls, Wireless VPN, and Wireless IDS. As we can conclude, it is obvious that although WLANs, as a viable supplement to wired LAN, have been widely accepted in our real life, it is still in its infant stages as long as security is concerned.

References

[1] Airfart. http://airfart.sourceforge.net/.

[2] AirSnort. http://airsnort.shmoo.com/.

[3] Airtraf. http://airtraf.sourceforge.net/.

[4] Cisco Wireless LAN Solution Engine (WLSE). Available at http://www.cisco.com/en/US/products/sw/cscowork/ps3915/. Accessed 10 November, 2011.

[5] Evil Twin Attack. Available at http://www.redoracle.com/index.php?option=com_remository&Itemid=82&func=fileinfo&id=59. Accessed 10 November, 2011.

[6] IEEE 802 Standards. Available at http://en.wikipedia.org/wiki/IEEE_802. Accessed 10 November, 2011.

[7] Kismet. http://www.kismetwireless.net/. Accessed 10 November, 2011.

[8] Rogue access point detection: Automatically detect and manage wireless threats to your network. http://www.proxim.com. Accessed 10 November, 2011.

[9] WAPI implementation plan. http://www.wapia.org/files/Guide. Accessed 10 November, 2011.

[10] Wavelink. http://www.wavelink.com. Accessed 10 November, 2011.

[11] Wellenreiter. http://www.kismetwireless.net/. Accessed 10 November, 2011.

[12] Wepcrack. http://wepcrack.sourceforge.net/. Accessed 10 November, 2011.

[13] Wep tools. http://www.redoracle.com/index.php?option=com_remository&Itemid=82&func=fileinfo&id=59. Accessed 10 November, 2011.

[14] Wireless LAN. http://en.wikipedia.org/wiki/Wireless_LAN. Accessed 10 November, 2011.

[15] WLAN Authentication and Privacy Infrastructure. http://en.wikipedia.org/wiki/WLAN_Authentication_and_Privacy_Infrastructure. Accessed 10 November, 2011.

[16] GB 15629.11-2003. Information technology telecommunications and information exchange between systems local and metropolitan area networks specific requirements part 11: Wireless Lan Medium Access Control (MAC) and Physical Layer (PHY) specifications.

[17] GB 15629.11-2003-XG1-2006. Information technology telecommunications and information exchange between systems local and metropolitan area networks specific requirements part 11: Wireless Lan Medium Access Control (MAC) and Physical Layer (PHY) specifications amendment 1.

[18] SMAC 2.0 MAC Address Changer. http://www.klcconsulting.net/smac/. Accessed 10 November, 2011.

[19] AirDefense. Tired of Rogues? Solutions for Detecting and Eliminating Rogue Wireless Networks. White paper, http://wirelessnetworkchannel-asia.motorola.com/pdf/. Accessed 10 November, 2011.

[20] Bahl P, Chandra R, Padhye J, Ravindranath L, Singh M, Wolman A, Zill B (2006) Enhancing the security of corporate Wi-Fi networks using DAIR. In Proc. MobiSys'06.

[21] Baiamonte V, Papagiannaki K, Iannaccone G, Torino P (2007) Detecting 802.11 wireless hosts from remote passive observations. In Proc. IFIP/TC6 Networking.

[22] Beck M, Tews E (2009) Practical attacks against WEP and WPA. In Proceedings of the 2nd ACM Conference on Wireless Network Security (WiSec).

[23] Bittau A, Handley M, Lackey J (2006) The Final Nail in WEPs Coffin. In IEEE Symposium on Security and Privacy.

[24] Siemens Enterprise Communications. WLAN Security Today: Wireless more Secure than Wired. http://www.enterasys.com/company/literature/WLAN%20Security%20Today-Siemens%20whitepaper_EN.pdf. Accessed 10 November, 2011.

[25] Corbett C, Beyah R, Copeland J (2006) A passive approach to wireless NIC identification. In IEEE International Conference on Communications (ICC'06).

[26] Cui A, Stolfo S (2010) A Quantitative Analysis of the Insecurity of Embedded Network Devices: Results of a Wide-Area Scan. In Annual Computer Security Applications Conference (ACSAC'10).

[27] Fluhrer S, Mantin I, Shamir A (2001) Weaknesses in the Key Scheduling Algorithm of RC4. In Lecture Notes in Computer Science, 2259: 1C24.

[28] Han H, Sheng B, Tan C, Li Q, Lu S (2009) A Measurement Based Rogue AP DetectionScheme. In IEEE International Conference on Computer Communications (INFOCOM 2009).

[29] He C, Mitchel J (2005) Security Analysis and Improvements for IEEE 802.11i. In Proceedings of the Network and Distributed System Security Symposium (NDSS'05).

[30] Ilyas M, Ahson S (2005) Handbook of Wireless Local Area Networks: Applications, Technology, Security, and Standards (Internet and Communications). CRC Press.

[31] Jana S, Kasera S (2008) On Fast and Accurate Detection of Unauthorized Wireless Access Points Using Clock Skews. In The Annual International Conference on Mobile Computing and Networking (MobiCom08).

[32] Ma L, Teymorian A, Cheng X (2008) A hybrid rogue access point protection framework for commodity Wi-Fi networks. In Proc. IEEE INFOCOM 2008.

[33] Mano C, Blaich A, Liao Q, Jiang Y, Cieslak D, Salyers D, Striegel A (2008) RIPPS: Rogue identifying packet payload slicer detecting unauthorized wireless hosts through network traffic conditioning. ACM Transactions on Information and System Security (TISSEC), 11(2): 1 – 23.

[34] Moen V, Raddum H, Hole K (2004) Weaknesses in the Temporal Key Hash of WPA. In Mobile Computing and Communications Review, pp. 76C83.

[35] Moskowitz R (2003) Weakness in Passphrase Choice in WPA Interface. http://wifinetnews.com/archives/2003/11/weakness_in_passphrase_choice_in_wpa_interface.html. Accessed 10 November, 2011.

[36] Pervaiz M, Cardei M, Wu J (2008) Security in Wireless Local Area Networks, Proceedings of Y. Xiao and Y. Pan (eds.). Security in Distributed and Networking Systems, World Scientific Publishing Co Inc.

[37] Shetty S, Song M, Ma L (2007) Rogue access point detection by analyzing network traffic characteristics. In IEEE Military Communications Conference (MILCOM'07).

[38] Song Y, Yang C, Gu G (2010) Who Is Peeping at Your Passwords at Starbucks? — To Catch an Evil Twin Access Point. In Proceedings of the 40th Annual IEEE/IFIP International Conference on Dependable Systems and Networks (DSN'10).

[39] Srilasak S, Wongthavarawat K, Phonphoem A (2008) Integrated Wireless Rogue Access Point Detection and Counterattack System. In International Conference on Information Security and Assurance, pp. 326 – 331.

[40] Vladimirov A, Gavrilenko K, Mikhailovsky A (2008) Wi-Foo: The Secrets of Wireless Hacking. Addison-Wesley Professional, Boston.

[41] Walker J (2005) 802.11 Security Series Part II: The Temporal Key Integrity Protocol (TKIP). http://jcbserver.uwaterloo.ca/cs436/handouts/miscellaneous/Intel_Wireless_2.pdf. Accessed 10 November, 2011.

[42] Wei W, Jaiswal S, Kurose J, Towsley D (2006) Identifying 802.11 traffic from passive measurements using iterative Bayesian inference. In Proc. IEEE INFOCOM'06.

[43] Wei W, Suh K, Wang B, Gu Y, Kurose J, Towsley D (2007) Passive online rogue access point detection using sequential hypothesis testing with TCP ACK-pairs. In Proceedings of the 7th ACM SIGCOMM conference on Internet measurement (IMC'07).

[44] Wexler J (2010) 2009 Wireless LAN State-of-the-Market Report. http://www.webtorials.com/content/2010/03/2009-wlan-sotm.html. Accessed 10 November, 2011.

Chapter 4
Security in Wireless Metropolitan Area Networks

Lei Chen[1], Narasimha Shashidhar[2], Shengli Yuan[3], and Ming Yang[4]

Abstract

Wireless Metropolitan Area Networks (WMANs) provide wireless communications at acceptable bandwidth over much larger geographical areas compared to Wireless Local Area Networks (WLANs). Also known as Wireless Local Loop (WLL), WMANs are based on the IEEE 802.16 standards with commercial name Worldwide Interoperability for Microwave Access (WiMAX). With its global market growing eighty-five percent in 2010 to 1.7 billion U.S. dollars, WiMAX is becoming a major competitor among the prevailing wireless communications technologies. While improved IEEE 802.16 standards and amendments were published and adopted in almost every year of the past decade, existing standards still contain a number of security vulnerabilities inherent from deprecated versions. This chapter starts with an introduction to and overview of Metropolitan Area Networks (MANs), WMANs, WiMAX and IEEE 802.16 standards, then discusses the technical details of WiMAX and IEEE 802.16 security aspects such as confidentiality, integrity, key generation and management, as well as security vulnerabilities, treats, and countermeasures.

1 Department of Computer Science, Sam Houston State University, Huntsville, Texas 77341, USA.

2 Department of Computer Science, Sam Houston State University, Huntsville, Texas 77341, USA.

3 Department of Computer and Mathematical Sciences, University of Houston-Downtown, Houston, Texas 77002, USA.

4 Department of Information Technology, Southern Polytechnic State University, Marietta, Georgia 30060, USA.

4.1 Introduction

A Metropolitan Area Network (MAN), also referred to as Metro Network (MN), is a computer communication network over a city or a group of adjacent cities[1,2]. The primary motivation for implementing such networks is to enable an efficient transportation of data-oriented traffic in a much larger geographical area compared to WLANs[3]. One of the major challenges that need to be addressed before deploying a metro network is scalability. Metro Networks based on conventional technologies do not offer cost-effective scaling to achieve the high capacities demanded of these networks. As an example, researchers at Sprint's Applied Research & Advanced Technology Labs (AR&ATL) have proposed a next-generation, high-capacity metropolitan area network under their HORNET project that is designed to achieve cost-effective scaling using hybrid optoelectronic ring network[4]. An all-optical high-capacity network architecture capable of supporting several hundred to a thousand nodes has also been proposed by researchers at IBM[5].

As noted before, a metro network typically spans a large city or a large campus, providing means to interconnect a number of local area networks using fiber-optic links or other high-capacity backbone technologies to offer uplinks to the Wide Area Networks (WANs) and the Internet. According to the IEEE 802-2002 standard[6], a LAN is generally owned and operated by a single organization whereas a MAN is typically designed to be used by many individuals and organizations, and sometimes also as public utilities. The terms LAN and MAN encompass a number of data communication technologies and applications that offer a wide range of services, including but not limited to, file transfer, graphics, text and data processing, email, database access and multimedia. The generally agreed upon consensus for classifying and distinguishing between LAN and MAN is the geographic region served: LANs typically are confined to a region of $0 \sim 2$ miles, and MANs span anywhere from 2 miles to 30 miles. Some technologies that are used in these networks are Asynchronous Transfer Mode (ATM), Fiber Distributed Data Interface (FDDI), Switched Multi-megabit Data Service (SMDS) and linked together using microwave, radio, infra-red or Ethernet-based connections.

Some common examples of MANs can be found in large cities where the fire stations and emergency responder networks are interlinked across jurisdictions. Media companies such as newspapers, cable networks employ metro networks to coordinate their activities across different branch offices. In the next few sections, we will discuss the dominant WMAN technology WiMAX, originally called Wireless MAN, and its security.

4.2 Fundamentals of WiMAX

The best-known WMAN technology is WiMAX, which can provide up to 70 Mbps of bandwidth over a radius of several miles. WiMAX is now being used by consumers in over 150 countries and gaining acceptance in several industries. The WiMAX Forum is an industry-led, not-for-profit organization that certifies and promotes the compatibility and interoperability of broadband wireless products based on the IEEE 802.16 Standards. The next subsection discusses WiMAX technologies in more details.

In this section, we will examine the fundamentals of WiMAX and the IEEE 802.16 Standards for this technology. The IEEE 802.16 working group was formed to address the projected increase in the demand for metropolitan and wide-area wireless internet access over the next few years. This working group has put forth a standard for Broadband Wireless Access (BWA) systems, namely the IEEE 802.16 Standards. In this section we will also discuss the applications, technical aspects of the standard, and the services that can be expected by the end users. We note that this section is devoted to the IEEE 802.16 Standards and does not cover the IEEE 802.20 or 802.22 Standards which seek to extend the IEEE 802.16 Standards.

4.2.1 WiMAX overview

In June of 2001, a number of technology corporations and service providers came together to establish the WiMAX Forum. The forum was created with the objective of accelerating wide-scale adoption of the Broadband Wireless Access technology. This forum had three primary goals in mind[7]:

- to establish standards and build profiles for equipment that ensures interoperability;
- to work with government agencies to release spectrum;
- to establish and grow an ecosystem nurturing vendor innovation and carrier deployment to encourage mass adoption of WiMAX technologies.

The success of the WiMAX Forum can be measured from the fact that in addition to attracting the traditional communications industry, partners of the WiMAX forum now include industries ranging from aviation, education, and energy to government and healthcare.

4.2.2 WiMAX network topologies

Four different network topologies are supported by the current IEEE 802.16 Standards: Point-to-Point (P2P), Point-to-Multiple-Point (PMP), Multi-Hop Relay (MHR), and Mobile[8].

The P2P topology involves a Base Station (BS), a Subscriber Station

(SS), and a dedicated long-range, e.g. up to 48 km (30 miles) using line of sight (LOS) or 8 km (5 miles) using None-LOS (NLOS), and high-capacity wireless link between the two parties.

The PMP adds more SSs to P2P topology which is commonly used to provide last-mile broadband access. Due to the cell configuration and the high density of obstacles and high interference to signals, the operating range reduces to less than 8 km (5 miles).

The MHR topology was defined in IEEE 802.16j-2009 and it extends the network by allowing SSs to relay traffic by acting as Relay Stations (RSs). The operating range between two nodes is less than 8 km (5 miles).

Similar to a cellular network, a Mobile topology WiMAX network has multiple BSs working together to provide seamless communications to SSs. The communication range is also within 8 km (5 miles).

4.2.3 The IEEE 802.16 Standards

The first IEEE 802.16 Standard, put forward by the IEEE Standards Board in December 2001, was based on LOS technology in the 10-66 GHz spectrum. Due to the lack of support for NLOS operations, this standard was not suitable for lower frequency applications. To this end, from 2002 to 2004, the IEEE 802.16a through 802.16d standards were published to accommodate this requirement. These standards were intended for consumers to use as a replacement for the 802.11 WLAN standard. In 2005, an amendment to 802.16a/d standard was made and the IEEE 802.16e was released. This standard has been commercialized under the name WiMAX, or Worldwide Interoperability for Microwave Access, by the WiMAX Forum Industry Alliance[9]. The Forum promotes and certifies compatibility and interoperability of products based on the IEEE 802.16 Standards. Table 4.1[10] summarizes the different standards in the IEEE 802.16 family.

Note that the standard 802.16a/d in Table 4.1 are "fixed" implying stationary and nomadic use with limitation that end devices cannot move between base stations but with the provision that they can enter the network at different locations[10]. The IEEE 802.16e standard mitigates this limitation and addresses the mobility enabling Mobile Stations (MSs) to handover between base stations while communicating.

The key features of the IEEE 802.16 Standards include:

- broadband Wireless Access;
- up to 48 km (30 miles) in distance and up to 70 Mbps in bandwidth;
- data rate vs. distance trade off using adaptive modulation 64QAM to BPSK;
- offers NLOS operation;
- 1.5 to 28 MHz channels;
- hundreds of simultaneous sessions per channel;
- delivers >1 Mbps per user;

- both licensed and license-exempt spectrum;
- QoS for voice, video, and T1/E1, continuous and bursty traffic, and
- supports PMP and Mesh network models.

Table 4.1 Summary of the IEEE 802.16 Standards

Standards	802.16-2001	802.16a/802.16d	802.16e
Spectrum	10 to 63 GHz	< 11 GHz	< 6 GHz
Channel Conditions	Line-of-Sight only	Non-Line-of-Sight	Non-Line-of-Sight
Speed (bit rate)	32 to 134 Mbps	75 Mbps max, 20 MHz channelization	15 Mbps max, 5 MHz channelization
Modulation	QPSK 16QAM 64 QAM	OFDM 256 subcarrier QPSK 16QAM 64QAM	Same as 802.16a
Mobility	Fixed	Fixed	Pedestrian mobility, regional roaming
Channel Bandwidths	20, 25, and 28 MHz	Selectable between 1.25 and 20 MHz	Same as 802.16a with sub-channels
Typical Cell Radius	1 ∼ 3 miles	3 ∼ 5 miles (up to 30 miles, depending on tower height, antenna gain and transmit power)	1 ∼ 3 miles

Besides the above IEEE 802.16 Standards, the IEEE 802.16-2009 Standard consolidated a number of previous 802.16 standards and amendments from 2004 through 2008. However the security aspects of IEEE 802.16-2009 are same as IEEE 802.16e and therefore is not considered separately as far as security is concerned.

The IEEE 802.16 Standards set different security requirements for various types of connections. Two types of connections are essential in WiMAX: management connections and data transport connections. Management connections have three subtypes: basic, primary, and secondary. A basic connection is created for each Mobile Station (MS), or an SS with mobility, when it joins the network. This type of connection is used for short and urgent management messages. The primary connection is also created for each MS at the same time with the purpose for delay-tolerant management messages. The secondary management connection is used for IP encapsulated management messages such as DHCP and SNMP. Transport connections are set up as needed and they are used to carry user data.

In Section 4.3 we will discuss the technical details of how security goals are achieved in WiMAX and IEEE 802.16 Standards.

4.3 WiMAX security goals and solutions

The IEEE 802.16e standard for WiMAX specifies a set of security mechanisms to protect confidentiality of data and secret keys, preserve integrity of data and control messages, and provide secure authentication as well as secure key generation and management[11,12]. These security goals are mainly addressed and achieved in the Security Sub-layer of the IEEE 802.16 Protocol Stack as shown in Fig. 4.1 and are discussed in the succeeding subsections[11,13]. Acronyms used in the following discussions are listed and explained as follows.

- RSA: a public key cipher very widely used in many secure authentication and communication protocols.
- Security Association (SA): a set of cryptographic methods and associated keying material that contains information about cryptographic ciphers and keys used. Each SS establishes at least one SA in the initialization process[14].
- Extensible Authentication Protocol (EAP): defined in Request For Comments (RFC) 3748 and updated in RFC 5247, EAP is an authentication framework widely used in wireless networks and Point-to-Point connections[15]. For WiMAX there are many methods defined by RFCs and a number of vendor specific methods each of which has its own keying material and parameters[15].
- Privacy Key Management (PKM) protocol: PKM provides the secure distribution of keying data from the BS to the SS and the synchronization of keying data. PKM is also used by the BS to enforce conditional access to network services. PKM Version 2 (PKMv2) was defined with enhanced security features in the 802.16e amendment[16].

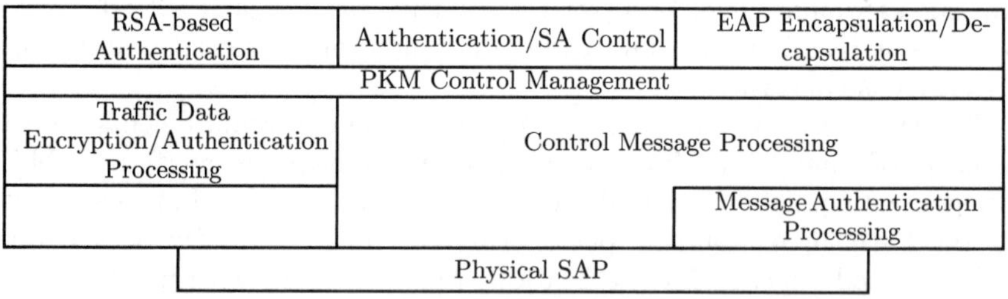

RSA-based Authentication
Authentication/SA Control
EAP Encapsulation/De-capsulation
PKM Control Management
Traffic Data Encryption/Authentication Processing
Control Message Processing
Message Authentication Processing
Physical SAP

Fig. 4.1 Security Sub-layer of IEEE 802.16 Protocol Stack.

The Security Sub-layer has been redefined in the IEEE 802.16e amendment to address a number of security vulnerabilities in the 802.16-2004 Standards. The 802.16e amendment defined PKMv2 with enhanced features. Table 4.2 summarizes the main differences between the two versions of PKM.

In the rest of this section, we will first discuss Digital Certificates (DCs) and Public Key Infrastructure (PKI) which aim to verify the binding of public key and the certified identity of the key owner. This is of great importance as the genuineness of the public key of communication parties determines the validity of subsequent key generation and exchange as well as other security processes. The discussion continues with Security Association followed by Key Generation and Management. A Security Association determines the security level and requirements of a connection. This section also discusses how authentication, confidentiality, and integrity are implemented in WiMAX with technical detail.

Table 4.2 The main differences between PKMv1 and PKMv2

Security Features	PKMv1	PKMv2
Authentication	RSA-based one-way authentication: the BS authenticates the SS	Mutual authentication. Supports two authentication methods: EAP or RSA
Security Association	One SA family: Unicast. Composed of three types of SAs: primary, dynamic and static.	Three SA families: Unicast SA, Group SA and MBS SA. Composed of the same three types of SAs as PKMv1
Key encryption	Use of three encryption algorithms: 3-DES, RSA and AES	New encryption method implemented: AES with Key Wrap
Data encryption	Two different algorithms are defined in the standard: DES in CBC mode, AES in CCM mode	Use of the same algorithms plus AES in CTR mode and AES in CBC mode implementation
Other additions	-	Management of security for: broadcast traffic, MBS traffic. Definition of a preauthentication procedure in the case of handover

4.3.1 WiMAX PKI and digital certificates

PKI is a set of hardware, software, people, policies, and procedures to create, manage, distribute, use, store, and revoke digital certificates[17,18]. The main purpose of PKI is to have trustworthy binding public keys with respective user identities by means of Certificate Authorities (CAs). In other words, if a certificate in the PKI is proven to be valid and genuine, the public key found in the certificate is then recognized as being genuine and the binding between this public key and the identity of the owner of the certificate is also validated (they are both from the same owner). According to the WiMAX PKI, each MS is preconfigured with an X.509 digital certificate[19]. The purpose of having X.509 certificates and their associated keys is to identify and authenticate the identity of devices or SSs and servers[20]. The WiMAX CAs provide hosting of the WiMAX PKI hierarchy and supplies device and server certificates for used in WiMAX networks.

WiMAX PKI makes use of public key cryptography, e.g. RSA, to digitally sign certificates in a hierarchy of certificates. As shown in Fig. 4.2, each certificate is digitally signed by a certificate at a higher level, back to a root certificate which signs itself to form a certificate chain. The format, content, and use of these X.509 certificates are described in RFC 3280. The cryptography related specifications, such as RSA cipher, can be found in the Public-Key Cryptography Standards (PKCS) #1 through #13 at RSA Laboratories website[17,20]. Also illustrated in the figure, WiMAX has two hierarchies of PKI, one for the device identification and the other for the identification of the Authentication, Authorization and Accounting (AAA) servers.

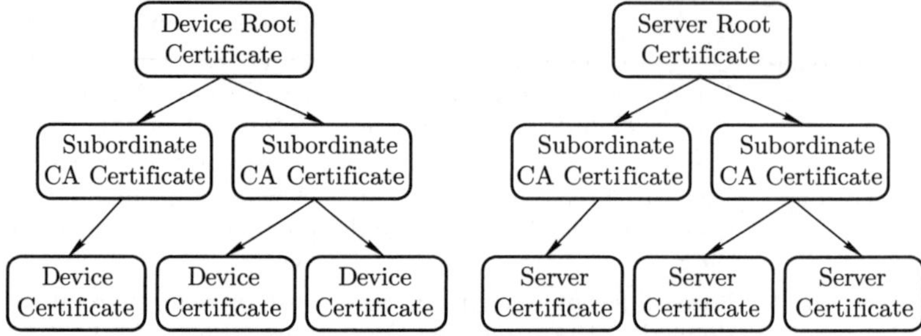

Fig. 4.2 Device and Server Hierarchies of WiMAX Public Key Infrastructure (PKI). Note: Arrows refer to the process of signing.

Figure 4.3 shows a segment of a certificate chain. In each of the certificates, regardless of a signing or signed certificate, the identity information, such as issuer identity and subject identity, the public key, and the validity dates are required[20]. The signature of certificate issuer is attached to the certificate. Among the above information, signature of issuer, issuer identity, and subject identity are used to distinguish between a signing certificate and a signed certificate, e.g. the signature of issuer for a signed certificate is obtained from the signing using the issuer's private key, and the subject's identity of a signing certificate appears as the issuer identity of all signed certificates.

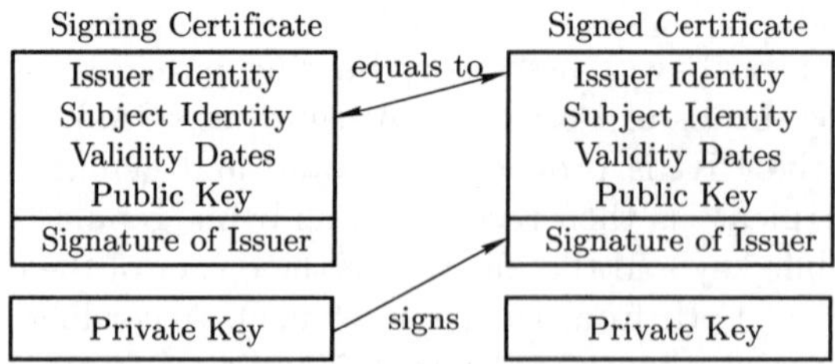

Fig. 4.3 A segment of a certificate chain.

The exchange of certificates and other authentication information is needed when a WiMAX SS tries to connect to a BS using Transport Layer Security (TLS)[20]. The SS sends the server its certificate chain which typically includes three to four certificates: its own certificate, the signing certificate, and all the higher signing certificates up to the root certificate, all of which are from the device hierarchy. On the other hand the AAA server sends its certificate chain, obtained from the server certificate hierarchy, to the SS. By accepting the digital certificates received, both the device and server accept the genuineness of the public keys included in the certificates, e.g. the public key found in a SS's certificate truly belongs to that SS. This process is essential in preparing for the key generation and management process as well as the authentication process.

4.3.2 WiMAX security association, key generation and management

Except for public keys, all other keys used in WiMAX are established during authorization, and are subject to an aging process. Consequently, they must be refreshed periodically during reauthorization. In WiMAX IEEE 802.16-2004, the following keys (not considering private keys as they are held only by the key owners) are defined in PKMv1[16,19].

- Public Key (PK): the PK is included in an X.509 digital certificate that belongs to the same owner. The genuineness and binding to identity is proven when the certificate is validated. An SS mainly uses its PK for authentication with the BS and encrypting the Authorization Key using R5A cipher.
- Authorization Key (AK): the BS determines the 160-bit AK and encrypts it with the MS's PK. After the MS receives the encrypted AK, it uses its private key to decrypt and obtain AK. The lifetime for an AK is between one and 70 days and the default value is 7 days. To provide smooth transitions, two AKs may be active at the same time and they are distinguished using sequence numbers (from 0 to 15).
- Key Encryption Key (KEK): a 128-bit KEK is determined by SS using AK. KEK is used as input to 3-DES cipher for encrypting the Traffic Encryption Key.
- Traffic Encryption Key (TEK): a 128-bit data encryption key with lifetime of 30 minutes to 7 days.
- Hashed Message Authentication Code Key (HMAC Key): the purpose of HMAC keys is to assure message integrity. There are three types of HMAC keys: 160-bit HMAC key for downlink (HMAC_KEY_D), 160-bit HMAC for uplink (HMAC_KEY_U), and HMAC key used in mesh mode (HMAC_KEY_S). MS uses AK to determine HMAC keys. The sequence number of AK implicitly affects the value of HMAC keys.

In additional to the keys discussed above, PKMv2 protocol in 802.16e amendment enhanced the security of Authorization Key generation and introduced a number of new keys mainly for multicast services as shown in Table 4.3[21].

WiMAX uses Security Association (SA) to specify the security parameters, such as keys and selected encryption algorithms, of a connection[19]. Among the four types of connections discussed previously, the basic and primary management connections do not have any associated SA and SAs are optional for secondary management connections but required for all transport connections. A transport connection can either have one SA for both uplink and downlink or two separate SAs for the two directions[19].

Table 4.3 New keys introduced in PKMv2 and the 802.16e amendment

Keys	Functions/Derivation
Pairwise Master Key (PMK)	obtained from EAP authentication
Primary Authorization Key (PAK)	obtained from RSA-based authentication
Authorization Key (AK)	derived from PMK or PAK
Group Key Encryption Key (GKEK)	used for encrypting GTEK
Group Traffic Encryption Key (GTEK)	used for encrypting multicast data packets
Multicast Broadcast Service (MBS) Authorization Key (MAK)	authentication for MBS
MBS Group Traffic Encryption Key (MGTEK)	used for generating MTK with MAK
MBS Traffic Key (MTK)	used for protecting MBS Traffic; derived from MAK and MGTEK
HMAC_KEY_D	used for preserving message integrity for downlink
HMAC_KEY_U	used for preserving message integrity for uplink

Each SA has a unique 16-bit identifier (SAID), a cryptographic suite identifier for selected algorithms, TEKs, and Initialization Vectors (IVs). SAs are managed by the Base Stations (BS). There are three types of SAs: primary SA, static SA and dynamic SA[19]. For each MS a primary SA is established for its secondary management connection with the BS and it is established when the MS is initialized. This primary SA is unique and is only shared between a specific MS and BS. During the initialization of an MS, the BS creates one or multiple static SAs depending on the services that MS has subscribed, e.g. there may be a static SA for the basic unicast service and additional static SAs for each of other subscribed services. Dynamic SAs are only created when there are new traffic flows and they are closed when traffic flows are completed. In scenarios such as multicast, static SAs and dynamic SAs can be shared among multiple SSs.

4.3.3 WiMAX authentication

The authentication in WiMAX includes three different types: BS authenticating SS as required in PKMv1, Mutual Authentication and Message Authentication required in PKMv2.

4.3.3.1 BS authenticating SS in PKMv1

In PKMv1, to start the authentication process, the SS sends the BS a PKM Authentication Information message which contains SS manufacturer X.509 certificate[16]. Then the SS sends a PKM Authorization Request message containing the same certificate, the SS primary SAID, and a description of its security capabilities, e.g. supported ciphers. Upon receiving the request message, the BS uses the validated Public Key found within the received X.509 certificate to encrypt the Authorization Key and sends it, along with other information such as AK lifetime, sequence number, and SA descriptor(s), to the SS using a PKM Authorization Response Message. Replay attacks can be prevented with the help from the above-mentioned sequence number. After this point, the SS is required to periodically repeat authentication and key exchange to keep its key material up-to-date. The above process only implements one-way (BS to SS) authentication and the opposite direction authentication is missing. This security vulnerability is addressed and fixed in PKMv2.

4.3.3.2 Mutual authentication in PKMv2

Although both versions of PKM share the same security basis[16], PKMv2 provides mutual authentication in which the SS also authenticates the BS. The mutual authentication process goes as follows.

- BS authenticates an SS.
- SS authenticates the BS.
- BS provides the authenticated SS with an AK.
- BS provides the authenticated SS with the identities and properties of primary and static SAs.

PKMv2 supports two different authentication protocols: X.509 digital certificates or Extensible Authentication Protocol (EAP). When X.509 certificate based authentication is used, the process is same in both directions as described in the previous subsection. If otherwise EAP is used, one of the defined and supported EAP authentication methods needs to be chosen, and corresponding security elements, such as subscriber identity module, password, X.509 certificate or others, will also be used in such method. Currently among the various available methods, Transport Layer Security (TLS) and Tunneled Transport Layer Security (TTLS) are recommended by the WiMAX Forum[22]. EAP-TLS is an Internet Engineering Task Force (IETF) open standard and is defined in RFC 5216[23]. It uses PKI to secure communication to a Remote Authentication Dial In User Service (RADIUS) or another type of authentication server[15]. EAP-TTLS extends EAP-TLS by providing a secure connection or "tunnel" for the BS to authenticate the SS[15].

4.3.3.3 Message authentication

In versions before 802.16e, the authentication of messages is done by using the Hashed Message Authentication Code (HMAC) which makes use of the HMAC keys discussed earlier. Since only the communication parties involved in the same SAs will share the same HMAC keys, only the authenticated parties are able to provide the correct HMAC. The original HMAC authentication did not provide a counter to protect against replay attacks and subsequently it has been fixed in newer versions. Another message authentication method, One-key Message Authentication Code (OMAC), is supported by 208.16e[19]. The OMAC is Advanced Encryption Standard (AES)-based and it includes replay attack protection. As far as message authentication is concerned, both HMAC with anti-replay counter and OMAC are considered to have strong security.

4.3.4 WiMAX confidentiality

In this section, we discuss how WiMAX protects the confidentiality of both keys and data.

4.3.4.1 Key confidentiality

The BS generates Authorization Key and encrypts it using RSA cipher with SS's PK. No other party, except the SS who possesses the paired private key, is able to decrypt and obtain AK. Both KEK and HMAC Keys are derived on both sides using AK[16,24]. During their lifetime, these keys must be stored in a secure manner at both SS and BS. The TEK for encrypting data is encrypted using one of the following ciphers: 3DES with 112-bit KEK, AES with 128-bit KEK, or RSA with SS's PK[24]. In PKMv2, the additional secret keys used for multicast and broadcast are also secured in a similar way.

4.3.4.2 Data confidentiality

The PKMv1 only uses Data Encryption Standard (DES) algorithm in the Cipher Block Chaining (CBC), or DES-CBC, to protect data confidentiality and PKMv2 adds AES-CCM with the active TEK[8,16,19].

The Data Encryption Standard (DES) is a symmetric-key block cipher using 56-bit key and was selected by the National Bureau of Standards (NBS) as an official Federal Information Processing Standard (FIPS) for the U.S. in 1976[25]. Mainly due to its small key size DES was found vulnerable about a decade ago. However, the algorithm is believed to be practically secure in the form of 3DES. In the Cipher-Block Chaining (CBC) mode, each plaintext block is XORed with the previous ciphertext block before being encrypted, as shown in Fig. 4.4.

The Advanced Encryption Standard (AES) is a symmetric key block cipher announced by NIST as U.S. FIPS in 2001 superseding DES[26]. The

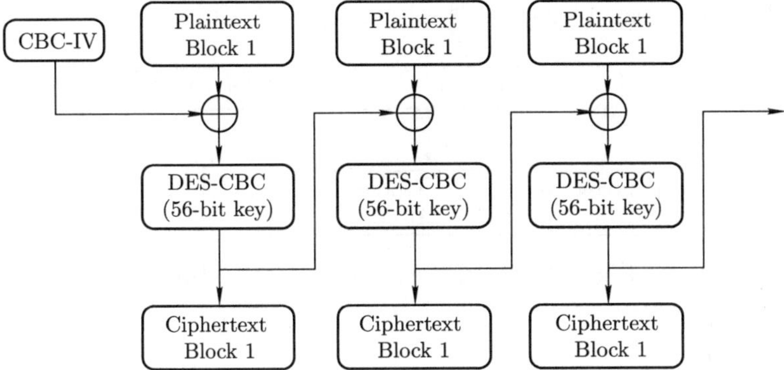

Fig. 4.4 Operations of DES-CBC.

Counter with CBC-MAC (CCM) mode of a cipher combines the Counter mode with CBC-MAC in order to provide both authentication and confidentiality. Counter mode generates the next keystream block by encrypting successive values of a "counter", e.g. keystream block 1 is generated by encrypting counter "00000000" with a nonce, and the next keystream block is generated by encryption counter "00000001" with the same nonce. Although increment counter is most frequently used, counter function can be any function that produces a sequence that will not repeat for a long time[27]. CBC-MAC stands for Cipher-Block Chaining Message Authentication Code, which constructs a MAC from a block cipher[28]. AES in CCM Mode in PKMv1 is considered a stronger cipher with longer key compared to DES-CBC and therefore is preferred in protecting data confidentiality.

In additional to the above ciphers, the 802.16e Amendment added a number of cryptographic algorithms, including AES in Counter for MBS, AES in CBC mode, and AES Key Wrap with 128-bit key[16]. The Key Wrap combines both encryption/decryption with integrity check values and can therefore provide both confidentiality and integrity.

4.3.5 WiMAX integrity

WiMAX preserves data integrity by using HMAC and Cipher-based Message Authentication Code (CMAC)[16,24], e.g. PKMv1 supports the use of HMAC for both downlink and uplink traffic and PKMv2 adds CMAC for the same purpose. The generation of HMAC or CMAC is illustrated in Fig. 4.5 where the Secret Key refers to the integrity related keys introduced earlier: HMAC_KEY_D, HMAC_KEY_U, and HMAC_KEY_S in PKMv1, and H/CMAC/KEY_D and H/CMAC/KEY_U in PKMv2. As addressed in the previous section, the AES Key Wrap in PKMv2 also provides integrity check of the traffic. When EAP exchange happens, the EAP messages are protected by an EAP Integrity Key (EIK)[8,16].

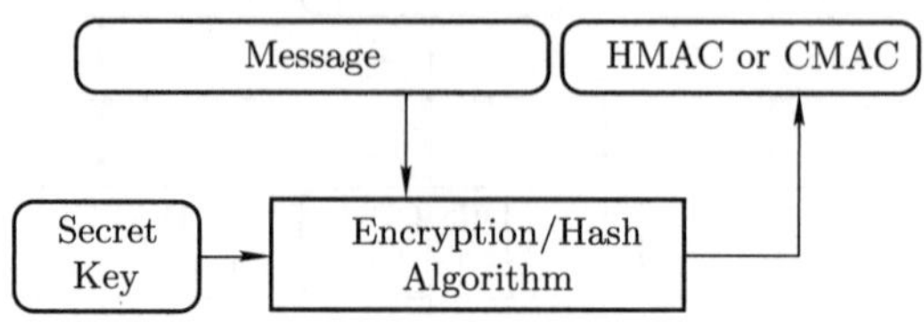

Fig. 4.5 Generation of HMAC or CMAC.

4.4 WiMAX security vulnerabilities, threats, and countermeasures

In this section, WiMAX security vulnerabilities, threats, and countermeasures will be discussed.

4.4.1 IEEE 802.16-2004 WiMAX systems

In WiMAX systems that use versions prior to IEEE 802.16e, the major security vulnerabilities include[8,29]:

- No two-way authentication: not until 802.16e did WiMAX provide authentication of BS by SS. Therefore in earlier versions, SSs were susceptible to forgery attacks by a rogue BS. Threats making use of this vulnerability include degraded performance, information theft, Denial of Service (DoS) attacks, and Man-in-the-Middle (MiM) attacks. External authentication of devices and users should be used to identify the BS and enforce two-way authentication.
- Weak cryptographic algorithm: only DES-CBC was available for protecting data confidentiality. This potentially leads to unauthorized disclosure of information, eavesdropping, DoS attacks, and MiM attacks. Stronger cipher algorithms, e.g. FIPS-validated algorithms such as AES, should be employed.
- Reused TEK: due to the short identifier (only two bits in length) of TEK, it repeats every four rekey cycles. Threats include reusing expired TEKs in replay attacks to disclose confidential information and further compromise the TEK. It is recommended to use FIPS-validated encryption algorithms as well as cryptographic modules.

4.4.2 All WiMAX systems

The following vulnerabilities appear in all WiMAX systems regardless of the versions of the standards or amendments adopted[8,29,30]:

- Subject to RF jamming attacks: this is not unique to WiMAX as all wireless technologies are subject to such attacks. Classified as a DoS attack, RF jamming adversary transmits powerful RF signals to overwhelm the WiMAX spectrum causing all SSs within the interference range not being able to communicate. While it is possible to locate and remove the source of the RF jamming, this is often not an easy task considering the relatively large area covered by WiMAX, e.g. radius of 5 miles. Therefore out-of-band communications are recommended.
- Subject to scrambling attacks: while considered as a subcategory of RF jamming attacks, scrambling requires more precise injections of RF interference during the transmission of specific management messages in relatively short time periods and therefore is more difficulty to detect. Countermeasures to such attacks are similar to jamming attacks but require more sensitive and accurate detection and faster responses.
- Unencrypted management messages and no integrity check for multicast and broadcast traffic: none of the WiMAX standards or amendments so far has addressed or required the encryption of management messages, and consequently puts confidential information involved in the processes of network entry, node registration, and bandwidth allocation in danger. Possible related attacks include eavesdropping, replay attacks, and scrambling. Integrity checks are only provided to unicast traffic, leaving multicast and broadcast traffics are subject to DoS attacks. There is no countermeasure to this threat given that no encryption is applied to management messages. AES-CCM however helps in fighting against MiM attacks. As far as DoS attacks are concerned, it is recommended to plan for out-of-band communications and the inclusion of incident responses.

4.5 Summary

In this chapter, we discussed the basics of MANs, WMANs, WiMAX and its security elements, security goals and solutions, as well as vulnerabilities, threats, and countermeasures. Recent IEEE 802.16 Standards and WiMAX provide Mutual Authentication and Message Authentication, Data and Key Confidentiality, and Data Integrity. Though WiMAX is still subject to multiple threats and attacks, its security will continue to be strengthened in forthcoming IEEE 802.16 Standards and amendments. Being a strong competitor to Long Term Evolution (LTE) in 4G cellular networks, it will continue to provide service subscribers with relatively secure communications at satisfactory bandwidth over much larger geographical areas compared to

WLANs.

References

[1] Ghosh A, Wolter D R, Andrews J G, Chen R (2005) Broadband wireless access with wimax/802.16: current performance benchmarks and future potential. Communications Magazine, IEEE, 2005, 43(2): 129 – 136.

[2] MAN (2011) Retrieved from http://en.wikipedia.org/wiki/Metropolitan_area_network. Accessed 11 October, 2011.

[3] WiMAX (2011) Retrieved from http://en.wikipedia.org/wiki/WiMAX. Accessed 11 October, 2011.

[4] White I M, Rogge M S, Shrikhande K, Kazovsky L G (2003) A summary of the hornet project: A next-generation metropolitan area network. Selected Areas in Communications, IEEE Journal, 21(9): 1478 – 1494.

[5] Green P E Jr, Coldren L A, Johnson K M, Lewis J G, Miller C M, Morrison J F, Olshansky R, Ramaswami R, Smithand E H Jr (1993) All-optical packet-switched metropolitan-area network proposal. Lightwave Technology, 11(5): 754 – 763.

[6] IEEE LAN/MAN Standards Committee (2002) IEEE standard for local and metropolitan area networks: Overview and architecture. IEEE Standards, IEEE Std 802 – 2001.

[7] WiMAX Forum Goals (2011) Retrieved from http://www.wimaxforum.org/news/2839. Accessed 11 October, 2011.

[8] Guide to Securing WiMAX Wireless Communications (2011). Retrieved from http://csrc.nist. gov/publications/nistpubs/800-127/sp800-127.pdf. Accessed 11 November, 2011.

[9] WiMAX Forum (2011) Retrieved from http://www.wimaxforum.org/. Accessed 11 November, 2011.

[10] Nguyen T (2011) A survey of wimax security threats. Retrieved from http://www.cs.Wustl. edu/~jain/cse571-09/ftp/wimax2.pdf. Accessed 11 November, 2011.

[11] Nasreldin M, Asian H, El-Hennawy M, El-Hennawy A (2008) WiMAX Security. Advanced Information Networking and Applications – Workshops, 2008 (AINAW 2008). 22nd International Conference, pp. 1335 – 1340.

[12] Maccari L, Paoli M, Fantacci R (2007) Security Analysis of IEEE 802.16. Communications (ICC '07). IEEE International Conference, pp. 1160 – 1165.

[13] Yang E (2011) A Survey of WiMAX and Mobile Broadband Security. Retrieved from http://www.cs.wustl.edu/~jain/cse571-09/ftp/WiMAX1.pdf. Accessed 11 November, 2011.

[14] Eklund C, Marks R B, Stanwood K L, Wang S (2002) IEEE standard 802.16: a technical overview of the Wireless MANTM air interface for broadband wireless access. Communications Magazine, IEEE, 40(6): 98 – 107.

[15] Extensible Authentication Protocol (EAP) (2011) Retrieved from http://en.wikipedia.org/ wiki/Extensible_Authentication_Protocol. Accessed 11 November, 2011.

[16] Nuaymi L (2007) WiMAX: Technology for Broadband Wireless Access. John Wiley and Sons, New York.

[17] Public-Key Cryptography Standards (PKCS) (2011) Retrieved from http://www.rsa.com/ rsalabs/node.asp?id=2124. Accessed 11 November, 2011.

[18] Public Key Infrastructure (PKI) (2011) Retrieved from http://en.wikipedia.org/wiki/Public_ key_infrastructure. Accessed 11 November, 2011.

[19] Barbeau M, Laurendeau C (2007) Analysis of Threats to WiMAX/802.16 Security. Mobile WiMAX: Toward Broadband Wireless Metropolitan Area Networks. Wireless Networks and Mobile Communications Series, Volume 7, Taylor and Francis Group CRC Press, Boca Raton.

[20] WiMAX Public Key Infrastructure (PKI) Users Overview (2011). Retrieved from http://www.wimaxforum.org/sites/wimaxforum.org/files/page/2009/12/wimax_pki_users_overview_4_28_10.pdf. Accessed 11 November, 2011.

[21] Xu S, Matthews M, Huang T C (2006) Security issues in privacy and key management protocols of IEEE 802.16. In Proceedings of the 44th annual Southeast regional conference (ACM-SE 44), pp. 113 – 118.

[22] New EAP Variants (2011) Retrieved from http://www.wimaxforum.org/news/731. Accessed 11 November, 2011.

[23] RFC 5216: The EAP-TLS Authentication Protocol (2011) Retrieved from http://tools.ietf. org/html/rfc5216. Accessed 11 November, 2011.

[24] IEEE 802.16 WiMAX Security (2011) Retrieved from http://www.first.org/conference/2005/papers/kitti-wongthavarawat-slides-1.pdf. Accessed 11 November, 2011.

[25] Data Encryption Standard (DES) (2011) Retrieved from http://en.wikipedia.org/wiki/Data_Encryption_Standard. Accessed 11 November, 2011.

[26] Advanced Data Encryption (AES) (2011) Retrieved from http://en.wikipedia.org/wiki/Advanced_Encryption_Standard. Accessed 11 November, 2011.

[27] Counter Mode (2011) Retrieved from http://en.wikipedia.org/wiki/Block_cipher_modes_of_operation#Counter_.28CTR.29. Accessed 11 November, 2011.

[28] CBC-MAC (2011) Retrieved from http://en.wikipedia.org/wiki/CBC-MAC. Accessed 11 November, 2011.

[29] Barbeau M (2005) WiMAX/802.16 threat analysis. In Proceedings of the 1st ACM international workshop on Quality of service & security in wireless and mobile networks (Q2SWinet '05), pp. 8 – 15.

[30] Dekleva S, Shim F P, Varshney U, Knoerzer G (2007) Evolution and emerging issues in mobile wireless networks. Communication of the ACM — Smart business networks, 50(6): 38 – 43.

Chapter 5
Security in Bluetooth Networks and Communications

Lei Chen[1], Peter Cooper, and Qingzhong Liu

Abstract

This chapter is concerned with security management in Bluetooth communication. The chapter begins with description of the development of Bluetooth and its technical specifications. It continues with a discussion of the various network structures that can be developed through Bluetooth. A discussion of Bluetooth security goals leads to a description of the different security models available. We conclude with a discussion of the more prevalent attacks on Bluetooth security and the most widely used procedures for mitigating such attacks.

5.1 Introduction

Bluetooth is relatively a new technology to provide short distance wireless communications[1−3]. Bluetooth does not depend on other types of networks such as the Internet or Local Area Networks (LANs) and it supports simple and instant connections and data exchanges between almost all kinds of devices installed with Bluetooth hardware and software modules. Bluetooth devices include smartphones, headsets, media players, vehicular devices, digital cameras, TVs, and computers.

In 1994, a team of researchers at Ericsson Mobile Communications initiated a feasibility study of universal short-range and low-power wireless connectivity mainly for exchanging data between mobile phones, headsets, vehicular devices, and computers[2]. Four years later a group named Bluetooth Special Interest Group (SIG) was formed with founding members from Er-

1 Computer Science Department, Sam Houston State University, Huntsville, Texas 77341, USA.

icsson, Nokia, Intel, IBM, and Toshiba[3]. Nowadays, the Bluetooth SIG has more than fifteen thousand member companies in the areas of telecommunication, networking, computing, and consumer electronics[4], and Bluetooth modules are integrated in most smartphone, wireless headsets, and many newly released notebook computers and vehicles.

The current Bluetooth technology operates over the 2.4 GHz Industrial-Scientific-Medical (ISM) unlicensed frequency band which is mainly for low-power transmissions. Within 10 to 100 meters, Bluetooth can provide 700 Kbps, 2.1 Mbps, or up to 24 Mbps data rates depending on its version[3]. Security in Bluetooth is provided through authentication, encryption, and exchanging keys in a secure manner. Bluetooth is not based on IP and does not make use of the more advanced and IP-based standard security features, such as Secure Socket Layer/Transport Layer Security (SSL/TLS), digital certificates, or IPSec. As a result it has and will continue to have security vulnerabilities and potential threats. Bluetooth was designed as a low-powered and low-cost communication system. These limitations produce challenges in the implementation resource intensive security strategies.

The rest of this chapter is organized as follows: Section 5.2 provides an introduction to the basic technical specifications, such as radio frequency, power control, communication range, data rates, and versions, as well as Bluetooth network architecture. Section 5.3 discusses the security of Bluetooth in subsections, include Security Goals, Security Modes, Key Generation and Management, Authentication, Confidentiality, Trust Levels, Service Levels, and Authorizations. Section 5.4 addresses Bluetooth vulnerabilities, potential threats, attacks, and countermeasures. The last section concludes the entire chapter.

5.2 Bluetooth Primer

Bluetooth is a semi-open standard for communications within short-range and ad hoc networks, such as Wireless Personal Area Networks (WPANs)[5]. Its peer-to-peer (P2P), low-cost and low-power characteristics enable Bluetooth to form small-sized ad hoc networks — piconets[1] as discussed in Section 5.2.2. While a Bluetooth piconet can support up to 8 devices (one "master" and seven "slaves"), many Bluetooth applications only involve pairs of Bluetooth devices, for example, between Bluetooth enabled cellular phones and headsets. The network topology and node capabilities are closely tied to the specifications of Bluetooth technologies as discussed below.

5.2.1 Bluetooth Technical Specifications

1. *Frequency*

Along with IEEE 802.11b/g WLAN and many other technologies, Bluetooth operates in the 2.4 GHz ISM radio bands[1]. In order to reduce signal interference in such crowded segment of frequency spectrum, Bluetooth makes use of Frequency Hopping Spread Spectrum (FHSS) technology for its communication. Bluetooth operates over 79 different radio channels, and is capable of hopping 1,600 times per second for data and voice, and 3,200 times per second for signaling. The dwell time on each channel is thus 625μs. Given that 259μs are required to implement the frequency switch and control exchange Bluetooth can transmit data for 366μs during each frame. Bluetooth allows for 1-slot, 3-slot and 5-slot frames. This gives data transmission rates, assuming 1MHz bandwidth and 1bit/Hz of 366bps for the 1-slot frame, 1.616 Kbps for a 3-slot frame and 2.866 Kbps for a 5-slot frame. FHSS not only lowers the chance of Bluetooth signals being intervened by other signals, but also provides a limited level of transmission security by changing frequency constantly. This makes it a little more difficult for a malicious node to locate the exact frequency being used and consequently to eavesdrop communication data.

2. *Radio link power control and communication range*

Bluetooth devices are capable of measuring the Received Signal Strength (RSS) and accordingly notify their neighbors for increasing or decreasing transmission power. This technology is especially useful in small-sized mobile devices to conserve limited power and extend battery life. Communication range of Bluetooth devices can vary from 1 meter to 91 meters depending on the type of power management[1]. Although a Bluetooth device can adjust its transmission power, there are significant differences among the power levels of various devices. For example, a Class 1 type of device, e.g. AC-powered Bluetooth devices, is powered at 100mW and can reach as far as 91 meters, while Class 2 type battery powered devices have up to 9-meter communication range at power level of 2.5mW. Low power-consuming Class 3 type devices such as Bluetooth adapters can only talk to neighbors within 1 meter at 1mW power level. While power control is not considered as a security mechanism, it helps to reduce the chance of being attacked, as an adversary needs to be within the communication range to launch an attack.

3. *Data rates and versions*

The rate at which a Bluetooth device can transmit data depends on the version of Bluetooth standard it supports. For Bluetooth 1.1 and 1.2, the transmission rate can be up to 1 Mbps and for versions 2.0 + Enhanced Data Rate (EDR) and 2.1 + EDR, it may be as high as 3 Mbps. In general the throughput is around 70% of the corresponding data rate. Bluetooth 3.0 + High Speed (HS) and 4.0 both support "Bluetooth over Wi-Fi" reaching

24 Mbps data transmission rate. The throughput of this scenario however depends on the performance of the carrier Wi-Fi network[4].

5.2.2 Bluetooth Network Architecture

Bluetooth allows two types of networks: Ad Hoc and infrastructural. A Bluetooth Access Point (AP) facilitates the communication among connected Bluetooth devices in an infrastructural network, while in Ad Hoc networks Bluetooth devices establish direct connections without any intermediary. The Ad Hoc type of network is far more common than infrastructural.

A Bluetooth device can be divided into two functional parts: host and host controller[1]. The functionality of a host is implemented in the base device such as a computer to which the Bluetooth module is connected. Functions of the host include implementing the upper layer protocols such as Logical Link Control (LLC), and Adaptation Protocol (L2CAP) and Service Discovery Protocol (SDP). On the other hand, the host controller, normally installed in a USB dongle or integrated as an embedded module, is in charge of the lower layer functions such as signaling, Baseband, and Link Manager Protocol (LMP). In many handheld devices such as smartphones and even smaller units like Bluetooth headsets, host and host controller are integrated into one single unit.

1. *Infrastructural Bluetooth networks*

Infrastructural Bluetooth networks fit to scenarios where a geographically fixed Bluetooth Access Point (AP) is used as the centralized communication hub for other Bluetooth devices. As this Bluetooth architecture is not commonly used, we will only focus on Ad Hoc Bluetooth networks in this chapter.

2. *Ad Hoc Bluetooth networks*

Without fixed infrastructure, Ad Hoc Bluetooth networks can be formed anywhere and anytime when Bluetooth enabled devices are within communication range. As devices can have their own moving velocity and direction, the topology of such network can change dynamically.

3. *Bluetooth Piconets and Scatternets*

A piconet is a collection of up to eight active Bluetooth devices. One of the devices is designated as the "master" device. The remaining seven are "slave" devices. The slaves perform clock synchronization with the master device upon joining the piconet. Communication between master and slave(s) can be one-to-one or one-to-many. In other words, the master can communicate with a single slave, with a subset of slaves or all of them. In addition to the eight active devices a piconet can accommodate an additional eight devices in an inactive or "parked" state. Switch state from active to inactive and vice versa can occur dynamically. Bluetooth devices can form scatternets. A scatternet is a collection of piconets where a slave within one piconet takes on the role of the master of an adjacent piconet.

A piconet master uses Time Division Duplexing-Time Division Multiple Access (TDD-TDMA). TDD-TDMA is a half-duplex form of communication where the master uses even numbered hops and the slave uses odd numbered hops. In one-to-one communication the master divides the communication evenly. In one-to-many communication the master has access to 50% of the channel with the slaves sharing the remaining 50% in round robin fashion. A device serves as the master of one piconet in the current time slot can act as a slave of another in the next time slot. FHSS allows for the slave-turned-master to communicate to its piconet without interfering in the original piconet through the use of a different spectrum hop sequence. Scatternets are certainly possible but there are few implementations as a result of limitation of both the Bluetooth and MAC layer specifications.

While Bluetooth supports complex topology with scatternets, the following security discussions are mainly based on the much more popular applications for communications between paired devices.

5.2.3 The Bluetooth Controller Stack

The Bluetooth controller stack provides three modes of communication: Asynchronous Connection-oriented Linkage (ACL) primarily used for data transmission, Synchronous Connection-oriented Linkage (SCO), used for voice data, and Stream Link used for continuous data flow.

ACL can provide a variety of data packets that can be distinguished by length (1-slot, 3-slot, and 5-slot frames), by error correction method, and by the optional use of modulation to increase data transfer rates. An ACL packet is automatically transmitted if the packet remains unacknowledged. ACL can achieve a data transfer rate of up to 721 Kbps.

SCO makes use of an existing ACL link, reserving a set of timeslots for voice data. SCO does not provide for retransmission. eSCO (enhanced SCO) extends the SCO framework allowing retransmission, and a wider range of packet types. A slave device can manage up to three SCO channels, each delivering 64 Kbps.

Stream Link is used for continuous data transmission. The data is transmitted unframed and is broadcast to all available devices within range. Stream Link is typically simplex with the master pushing data one way.

The Link Management Protocol (LMP) layer is responsible for the setup, management, and termination of connections between a master and slave device. LMP monitors the communication channel, sends and receives control information, obtains data on device capabilities, and supervises the inactive timeout mechanism.

5.3 Bluetooth Security Solutions

In this section, we discuss Bluetooth security aspects and solutions in details. We first define the security goals to achieve in Bluetooth communications. We then identify four security modes, each with different security requirements, for various security levels that are appropriate for a variety of applications. The discussion examines Key Generation and Management and Authentication and Confidentiality. Bluetooth Trust Levels, Service Levels, and Authorizations are reviewed at the end of this section.

5.3.1 Bluetooth Security Goals

Bluetooth security aims to provide data confidentiality, device authentication, and authorization. Data confidentiality refers to preventing unauthorized viewing of confidential data. Authentication involves verifying the identity of Bluetooth device involved in communication. In contrast to other types of network, Bluetooth natively is not concerned with user authentication, e.g. whoever has access to Bluetooth enabled smartphone will be allowed to use the paired in-vehicle hands-free Bluetooth receiver. Authorization happens after successful authentication where the goal is to ensure that a device is authorized to make use of certain service. The technical details of how these security goals can be achieved are elaborated in the rest of this section.

5.3.2 Bluetooth Security Modes

In order to meet various security demands, four security modes[1,6] were designed and implemented and each Bluetooth device must operate in one of the following four security modes. Each Bluetooth version supports one or multiple (not all) security modes.

Security Mode 1 does not provide any sort of security and is only supported in version 2.0 + EDR and earlier. Mode 1 is also called promiscuous mode in which devices do not employ any security mechanism or prevent other devices from establishing connections. Consequently identification of partner devices is not verified and data is not encrypted.

In Security Mode 2, a service level-enforced security mode, authentication and encryption are implemented at the LMP layer and its procedures are initiated between LMP link establishment and L2CAP channel establishment. Supported by all Bluetooth devices, Security Mode 2 allows the security manager to determine whether access to a specific device should be granted. For this purpose, the security manager maintains policies for access control and interfaces with other protocols and device users. To facilitate the security requirements for different application operating simultaneously, multiple security policies and trust levels for restricting access need to be defined, e.g. in Security Mode 2 device A may have the access to device B while being blocked from access to device C at the same time.

Security Mode 3, supported by v2.0 + EDR devices, is a link level enforce security mode, which means that security procedures are initiated before the physical link is fully established. Authentication and encryption (using symmetric key encryption) for all connections in both directions is mandatory.

Introduced in, and mandatorily required by Bluetooth v2.1 + EDR, Security Mode 4 is similar to Security Mode 2 in the sense of it being a service level enforced security mode where security procedures are initiated after link setup. While the authentication and encryption algorithms remain unchanged, Elliptic Curve Diffie Hellman (ECDH) algorithm is used for Link Key generation and key exchange in the Secure Simple Pairing process. In this mode, security requirements are classified as authenticated Link Key required, unauthenticated Link Key required, or no security required. The Secure Simple Pairing Association model being used will determine which of the above security requirement is applied.

The security mode applied in securing communications between a pair of Bluetooth devices is dependent on the nature of the application and the design of the product. For example, data exchange between a smartphone and paired Bluetooth headset may require a security mode that provides data encryption and secure key exchange, while a vehicular Bluetooth receiver which is designed to be only paired with in-vehicle Bluetooth devices may not demand a secure service at all.

5.3.3 Bluetooth Key Generation and Management

The method for generating Link Key is same in Security Modes 2 and 3 and is different from the Link Key generation method used in Security Mode 4. The following two subsections discuss these two methods.

5.3.3.1 Link Key Generation in Security Modes 2 and 3

The Bluetooth pairing process starts with a secure PIN code, which can be any 16-byte UTF-8 string[7,8] but is typically as short as 4-decimal-digit[2,9]. The steps for generating encryption key, shared between devices for data protection are as follows and are illustrated in Fig. 5.1.

Step 1: A Bluetooth device (denoted as device 1), embedded in a vehicle for data streaming with in-vehicle Bluetooth mobile devices, is just powered on or its Bluetooth function is just enabled. Device 1 senses other Bluetooth-enabled devices in its communication range and finds a smartphone (denoted as device 2).

Step 2: Device 1 generates a random number IN_RAND and acquires PIN number (typically "0000") from user or application, then sends IN_RAND to

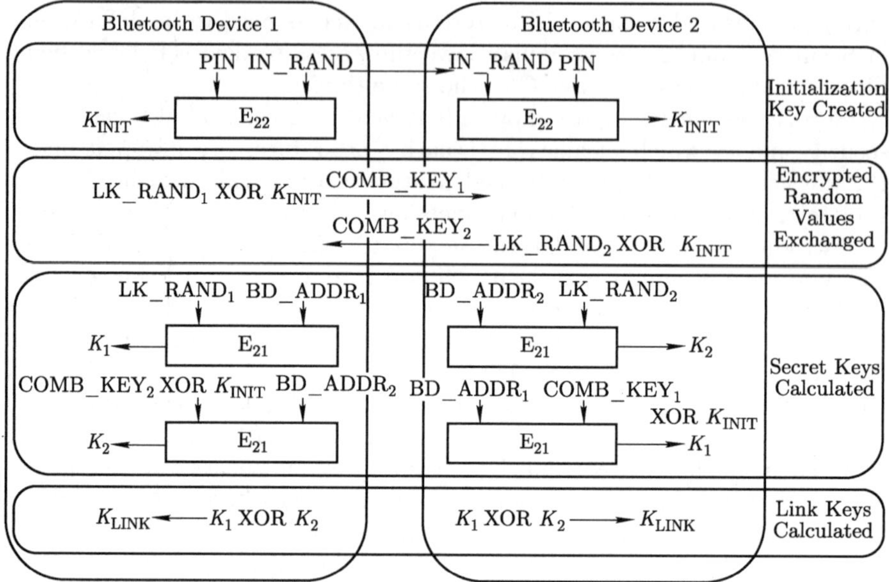

Fig. 5.1 Bluetooth key exchange in security modes 2 and 3.

device 2.
Step 3: Device 2 acquires the same PIN number (e.g. "0000") from user or application. Devices 1 and 2 both compute Initialization Key K_{INIT} using the PIN and random number IN_RAND as inputs.
Step 4: Device 1 computes Combination Key Component 1, COMB_KEY_1 using its local random number LK_RAND_1 and the Initialization Key K_{INIT}. Similarly Device 2 computes Combination Key Component 2, COMB_KEY_2 using its local random number LK_RAND_2 and the Initialization Key K_{INIT}. Then the two devices exchange these two Combination Key components.
Step 5: Locally both devices compute the two secret keys K_1 and K_2. More specifically device 1 computes K_1 using encryption algorithm E_{21} with its local random number LK_RAND_1 and address BD_ADDR_1. In order to obtain K_2, device 1 first XORs the received Combination Key Component 2 COMB_KEY_2 with the Initialization Key K_{INIT} to obtain LK_RAND_2 (refer to forthcoming discussions on logic behind this), then applies E_{21} with LK_RAND_2 and BD_ADDR_2 as inputs. Device 2 goes through the similar process to obtain both K_1 and K_2.
Step 6: A shared symmetric key K_{LINK} for data encryption is generated on both sides by simply XORing K_1 and K_2.

A number of issues in the above steps are further discussed below.

1. *Generation of Initialization Key*

The Initialization Key K_{INIT} is only used in the initialization process and must be discarded when the key exchange between two devices has completed.

K_{INIT} is derived using E_{22} algorithm with the PIN and a 128-bit random number IN_RAND[1] (and possibly the length of PIN "L" as well)[9,10].

2. *Generation and Exchange of Combination Key*

Each of the two Combination Key Components is generated on the two devices locally and then exchanged over the link[1,11]. The components are simply the result of XORing the local random number with the Initialization Key K_{INIT} they already agreed on. Note that none of the random numbers, LK_RAND_1 and LK_RAND_2, or the Initialization Key K_{INIT} is exchanged over the link.

3. *Calculating Secret Keys and Establishing the Link Key*

There is nothing special in the above steps except for how device 1 obtains K_2 and device 2 obtains K_1 respectively. Here the key computation makes use of a special property of the XOR operation: any arbitrary binary number XORing another arbitrary binary number twice will generate the same result as itself XORing zero and further returns itself. For device 1, this means:

$$\begin{aligned} &(\text{LK\_RAND}_2 \text{ XOR } K_{\text{INIT}}) \text{ XOR } K_{\text{INIT}} \\ &= \text{LK\_RAND}_2 \text{ XOR } (K_{\text{INIT}} \text{ XOR } K_{\text{INIT}}) \\ &= \text{LK\_RAND}_2 \text{ XOR } 0 = \text{LK\_RAND}_2 \end{aligned}$$

This is how device 1 obtains the value of device 2's local random number in Step 5 above. Similarly device 2 can obtain device 1's local random number LK_RAND_1. Now device 1 runs E_{21} to calculate K_2 using LK_RAND_2 and BD_ADDR_2 which is not secret. In a similar way device 2 can compute the Secret Key K_1. With the possession of both Secret Keys K_1 and K_2, both devices can calculate the Link Key K_{LINK} by simply XORing the two secret keys.

5.3.3.2 Link Key Generation in Security Mode 4

Instead of generating a shared secret symmetric key between Bluetooth devices, Security Mode 4 uses ECDH public/private key pairs and Secure Simple Pairing (SSP) which provides four different association models that fit into devices with various input and display capabilities. Security Mode 4 was introduced in v2.1 + EDR and fulfills three different security service requirements: authenticated Link Key, unauthenticated Link Key, and no security at all. Starting from v2.1 + EDR, Security Mode 1 (no security) and Security Mode 3 (link level security) are excluded and all devices are required to use Security Mode 4 except when pairing legacy devices which do not support such security mode and therefore have to use Security Mode 2[1,12]. The Link Key establishment and pairing process for SSP in Security Mode 4 is illustrated in Fig. 5.2 as well as discussed below[12].

An example of Bluetooth SIM (Subscriber Identity Module) Access Profile (SAP) is used here to discuss SSP process[12]. SAP defines the protocols and

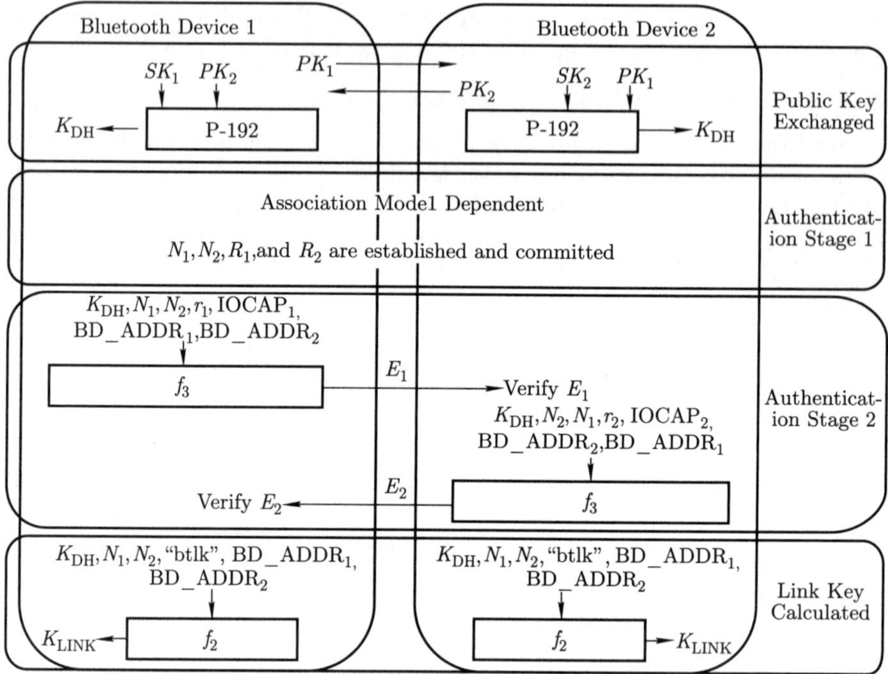

Fig. 5.2 Bluetooth link key establishment for secure simple pairing in security Mode 4.

procedures used to access a SIM card via a Bluetooth link[13].

Step 1 Device Discovery: In this step the SAP client looks for devices supporting SAP server. A Bluetooth Device Address (BD_ADDR) is required and can be obtained via a Bluetooth Inquiry. An Extended Inquiry Response (EIR) tag can help filter SAP server devices and so user can easily identify the known SAP server.

Step 2 Connection Establishment: The SAP server initiates authentication before initiating L2CAP channel establishment.

Step 3 IO Capability and Public Key Exchange: In order to determine which of the four association models should be used in pairing, IO capability on both the SAP client and server needs to be exchanged. Elliptic Curve Diffie-Hellman public keys from both sides are also exchanged. By the end of this step, both devices derive K_{DH} as shown in Fig. 5.2.

Step 4 Authentication Stage 1: Stage 1 varies slightly depending on which association model is being used. Regardless of association models, the purpose of this authentication stage is to display a six-digit number on a device and have the user to enter the same number on the pairing device.

Step 5 Authentication Stage 2: In this stage, the results of cryptographic functions are compared on both SAP client and server, and if they match both

devices will move on to calculate the Link Key.
Step 6 Link Key Calculation: A Link Key K_{LINK} is generated and mutual authentication is performed to ensure K_{LINK} is actually shared by both devices. Note that "btlk" in Fig. 5.2 is basically a string which is mapped to 32-bit Key ID using extended ASCII[14].
Step 7 Enable Encryption: As soon as Link Key has been established, the SAP client starts the L2CAP channel establishment procedure. This ends the key generation and exchange procedure.

The four association models are discussed below[1,12].

1. *Numeric Comparison*

Both devices using this model must be capable of displaying a six-digit number as well as allowing "yes" or "no" input from a user (e.g. between two smartphones). The six-digit number displayed is an output of the underlying security algorithm. During the process of pairing, a user is shown a six-digit number on each display and responds with a "yes" on each device when the numbers match. On the other hand, a respond of "no" will fail the pairing. A significant of this type of pairing compared with legacy pairing using PINs is that the displayed number is not used in the link key generation process and therefore an attacker capable of viewing the displayed number will not be able to use it for determining the link key.

2. *Passkey Entry*

This model is especially useful in scenarios such as between a PC and a Bluetooth enabled keyboard where the keyboard only has input capability whereas the PC has display capability. To make a pairing in this model, the six-digit number is shown on the device capable of displaying and exact same number must be entered on the other device, which is only capable of input. Similar to Numeric Comparison, the displayed number has nothing to do with generated link key. However, the security level provided in this model is still considered much higher compared to that of a legacy device with fixed PIN.

3. *Just Works*

In this association model, at least one of the pairing devices has neither display nor input capability (e.g. a Bluetooth enabled headset). The model executes Authentication Stage 1 just like Numeric Comparison model except for the appearance of a display. This model is not immune from Man In The Middle (MITM) attacks due to user being required to accept a connection without verifying the calculated value on both devices.

4. *Out of Band (OOB)*

In this model, pairing devices are capable of using a different wireless technology other than Bluetooth for device discovery and exchanging secrets to be used in pairing process[15]. The Near Field Communication (NFC)[16,17] is such an example where devices can be pair by simply tapping one against the other. Whether the model is subject to eavesdropping and MITM attacks depends on the security provided by the OOB used for pairing.

5.3.4 Bluetooth Authentication

The Bluetooth Authentication makes use of a challenge-response scheme for the verifier to identify the claimant[2,18]. Successful authentication indicates that the claimant possesses the shared Link Key. The authentication process is depicted in Fig. 5.3 and steps of this process go as follows.

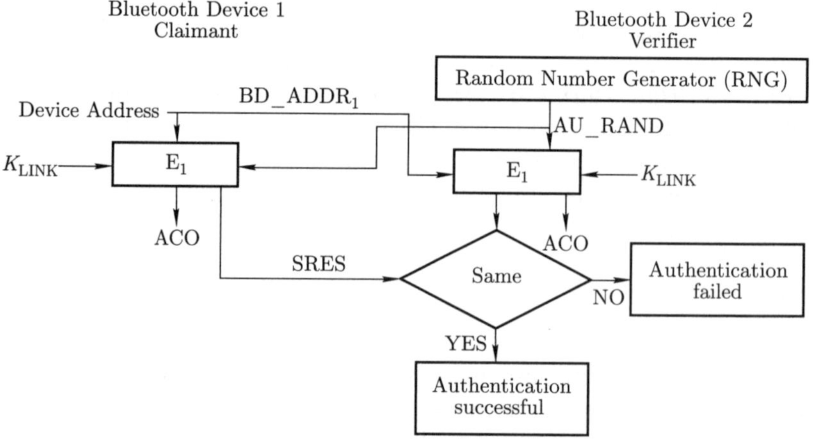

Fig. 5.3 Bluetooth authentication process.

Step 1: Verifier generates a 128-bit random number (AU_RAND) as challenge and sends it to the claimant.
Step 2: Both sides calculate the authentication response, using algorithm E_1 with claimant's Bluetooth device address (BD_ADDR), the challenge, and the shared Link Key. The output is 128-bit and the 32 most significant bits are sent from the claimant to the verifier as a response (SRES) and the remaining bits, known as Authenticated Ciphering Offset (ACO), will be used for creating the encryption key.
Step 3: The verifier compares the received SRES with the counterpart calculated locally.
Step 4: If they are same, authentication is successful; otherwise authentication fails.

Mutual authentication in Bluetooth is to perform the above authentication process again with the claimant and verifier swapped. The challenge AU_RAND must be different in every single authentication attempt to avoid replay attacks. Upon authentication failure, the claimant has to wait for an interval of time before the next authentication can start. This interval is increased exponentially against key exhaustive authentication attacks.

5.3.5 Bluetooth Confidentiality

Bluetooth provides not only the Security Modes discussed previously, it also offers a separate confidential service for better data confidentiality. Among the three Encryption Modes, Mode 1 requires no encryption, Mode 2 only requires individually addressed traffic to be encrypted using encryption keys based on individual Link Keys, and Mode 3 enforces encryption of all traffic using an encryption key based on the master Link Key.

As shown in Fig. 5.4, the operations in the upper half of the figure will generate the exact Keystreams on both the slave and master devices. The encryption on both directions is simply XORing the plaintext with the generated Keystream and decryption is to XOR the same Keystream to the ciphertext. In order to generate the Keystream, both sides use E_0 algorithm with the following inputs: random number EN_RAND, Master device's address BD_ADDR, clock, and the Constraint Encryption Key K_C which is derived using E_3 algorithm with the Link Key, EN_RAND and COF as inputs.

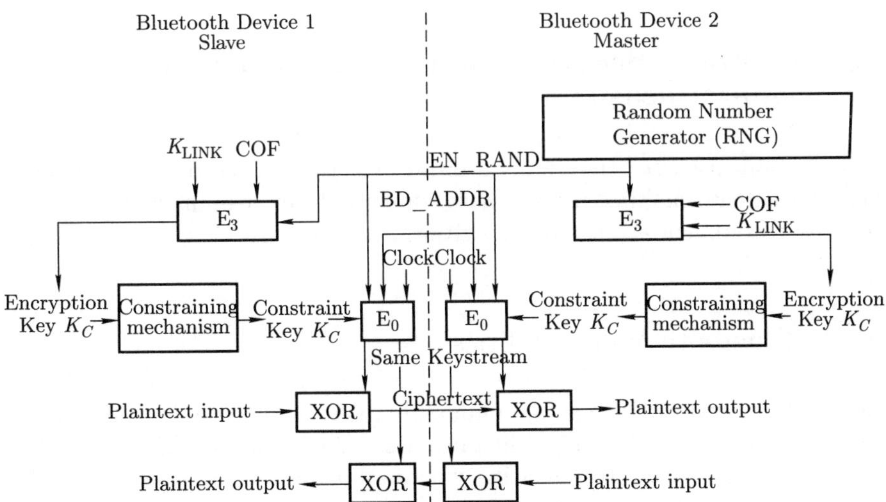

Fig. 5.4 Bluetooth Keystream generation and encryption procedure.

5.3.6 Bluetooth Trust Levels, Service Levels, and Authorization

In addition to the Security Modes, Bluetooth also provides two Trusted Levels as listed in Table 5.1.

Table 5.1 Bluetooth trust levels

Trust Levels	Device Relationship and Access to Services
Trusted	Fixed relationship and full access to all services
Untrusted	No relationship and restricted access to services

Three additional Security Service Levels are also available as listed in Table 5.2. Each different security requirement such as authentication, authorization, and encryption can be configured independently.

Table 5.2 Bluetooth security service levels

Service Levels	Authentication	Authorization	Access
Level 1	Required	Required	Automatic access only granted to trusted device
Level 2	Required	Not required	Access granted only after successful authentication
Level 3	Not required	Not required	Access granted automatically

5.4 Bluetooth Security Vulnerabilities, Threats, and Countermeasures

This section discusses the vulnerabilities of existing Bluetooth technologies and the threats making use of these vulnerabilities. Countermeasures and security recommendations are also given in this section.

5.4.1 Bluetooth Vulnerabilities

Before Bluetooth v1.2 the main vulnerability is related to the Unit Key which can be reused and becomes public once used. Consequently this may lead to eavesdropping of Unit Key sharing where an attacker may compromise the security between two other users if he or she has communicated with either or both of the two users using the Link (Unit) Key[1,19].

For Bluetooth versions before v2.1 the PIN becomes the main vulnerability. Although the PIN of a device can be as long as 16 bytes of UTF-8, most users tend to choose short PINs which are easy to brute force[8,20]. Also establishing PINs with a large number of users may be problematic as Bluetooth does not provide a PIN management mechanism. Another vulnerability is associated with the Encryption Keystream (as shown in Fig. 5.4), which will reappear due to the clock value (one of the inputs of E_0 algorithm for generating the Keystream) being repeated in a connection longer than 23.3 hours[1].

For all other versions of Bluetooth, vulnerabilities[1] are categorized and discussed as follows.

1. *Keys*

- Link Keys may not be properly stored.
- The allowed Encryption Key length can be as short as one byte which only generates a key space of 256 different keys.
- The Master Key is shared.

2. *Authentication*

- No user authentication — Bluetooth natively only provides authentication for devices. User authentication can be added in applications.
- Device authentication uses simple shared-key challenge-response.
- Although exponentially increased time intervals have been placed between authentication attempts, a designated limiting feature should be provided to prevent unlimited requests.

3. *Cryptographic Components*

- The Random Number Generator (RNG) used in Keystream generation (Fig. 5.4) may produce static number or periodic numbers.
- The E_0 stream cipher algorithm for encryption is weak[21,22].

4. *Privacy*

- Privacy, e.g. user activities, may be compromised if the Bluetooth Device Address (BD_ADDR) is captured and associated with a particular user.

5. *Security Services*

- End-to-end security service is unavailable.
- Bluetooth does not provide audit, nonrepudiation or other security services.

Besides the vulnerabilities listed above, a Bluetooth device should never be left in a discoverable or connectable mode as it is prone to attack in such status.

5.4.2 Bluetooth Threats and Countermeasures

Bluetooth is subject to a number of threats and attacks[23]. These attacks and their countermeasures[1] are discussed below.

- *Bluesnarfing*: This attack makes use of a flaw in the firmware of legacy devices. By forcing a connection to the device, such an attack can gain access to stored data and the International Mobile Equipment Identify (IMEI) which may lead to rerouting incoming calls to attacker's device[1,24,25].

The recommended countermeasure is to first update Bluetooth devices with an up-to-date operating system and software. Consider updating the hardware if it is generic and does not support the current security standards. It is also important to keep Bluetooth devices in non-discoverable mode anytime they are not actively exchanging data.

- *Bluejacking*: Similar to email spam and phishing, Bluejacking attacks send unsolicited messages to a Bluetooth enabled device to lure the user to conduct activities such as adding an entry in contact list[1,7,25]. Countermeasures include turning off Bluetooth device in certain public areas, such as shopping centers, when they are not being used and setting the device to hidden, invisible or non-discoverable mode from menu. If suspicious messages, typically coming from an admirer, a jokester or someone sending a business card, are found, simply ignore them by refusing or deleting them.
- *Bluebugging*: In this attack, devices and commands can be accessed through exploiting a flaw in firmware of legacy devices. Users are unaware of the existence of such attacks, which may access data, calls and other services. Countermeasures include updating both hardware and software of Bluetooth devices and requiring authentication.
- *Denial of Service* (*DoS*): Similar to a DoS in other types of wireless communication, such attacks overwhelm the target by sending large number of messages which freeze a device or drain the battery. Due to the short communication range, user detecting such attack can simply carry the device out of the danger zone. The best way to defend against DoS attacks is to limit device discoverability and connectivity by turning it off or hiding it in undiscoverable mode.

5.5 Conclusion

In this chapter, we first introduced the technical specifications and network architecture of Bluetooth as factors such as frequency, communication range, and connection type tie closely to security of communication. By clearly stating the security goals to achieve in Bluetooth technology, we have discussed the details of security aspects and techniques, such as Key Generation and Management, Authentication, Confidentiality, and Authority which are provided in different security modes implemented and supported by various Bluetooth versions. The intrinsic security pros and cons determine what kinds of vulnerabilities Bluetooth has and the potential threats and attacks it may face as well as the available countermeasures.

References

[1] Scarfone K, Padgette J (2008) Guide to Bluetooth Security: Recommendations of the National Institute of Standards and Technology. Special Publication 800-121, National Institute of Standards and Technology (NIST), U.S. Department of Commerce.

[2] Vainio J (2000) Bluetooth Security. Helsinki University of Technology. http://www.cse.tkk.fi/fi/opinnot/T-110.5190/2000/bluetooth_security/bluesec.html. Accessed 16 June, 2011.

[3] Bialoglowy M (2011) Bluetooth Security Review, Part I. Retrieved from http://www.syman tec.com/connect/articles/bluetooth-security-review-part-1. Accessed, 19 June, 2011.

[4] Bluetooth at Wikipedia (2011) Retrieved from http://en.wikipedia.org/wiki/Bluetooth. Accessed 15 August, 2011.

[5] Bluetooth Special Interest Group (2011) Specification of the Bluetooth System (Core Package version 4). Retrieved from http://www.bluetooth.com. Accessed 6 November, 2011.

[6] Gehramnn C (2011) Bluetooth Security White Paper. Bluetooth SIG Security Expert Group. Retrieved from http://grouper.ieee.org/groups/1451/5/Comparison%20of%20PHY/Bluetooth_24Security_Paper.pdf. Accessed 19 June, 2011.

[7] Bluetooth Security (2011) Retrieved from http://en.wikipedia.org/wiki/Bluetooth#Security from Wikipedia. Accessed 19 June, 2011.

[8] Shaked Y, Wool A (2005) Cracking the Bluetooth PIN. 3rd USENIX/ACM conf. Mobile Systems, Applications, and Services (MobiSys), pp. 39 – 50.

[9] Björnsson M (2001) Retrieved from http://www.lysator.liu.se/~martinb/Bluetooth/Bluetooth _in_Secure_Products.html#Toc520619560. Accessed 19 June, 2011.

[10] El-Hadidi M, Sayegh A (2005) A Modified Secure Remote Password (SRP) Protocol for Key Initialization and Exchange in Bluetooth Systems. International Conference on Security and Privacy for Emerging Areas in Communications Network, pp. 261 – 269.

[11] Jasobsson M, Wetzel S (2001) Security Weaknesses in Bluetooth. CT-RSA'01. LNCS 2020: 176 – 191. Springer-Verlag, Berlin.

[12] Bakshi A (2007) Bluetooth Secure Simple Pairing. Retrieved from http://www.wireless designmag.com/PDFs/2007/1207/wd712_coverstory.pdf. Accessed 16 October, 2011.

[13] SIM Access Profile (SAP) (2011) Retrieved from http://www.palowireless.com/infotooth/ tutorial /n12_sap.asp. Accessed 16 October, 2011.

[14] Simple Pairing Whitepaper (2011) Retrieved from http://mclean-linsky.net/joel/cv/Simple% 20Pairing_WP_V10r00.pdf. Accessed 16 October, 2011.

[15] Pasanen S, Toivanen P, Haataja K, Paivinen N (2010) New Efficient RF Fingerprint-Based Security Solution for Bluetooth Secure Simple Pairing. Hawaii International Conference on System Sciences, pp. 1 – 8.

[16] Near Field Communication (2011) Retrieved from http://en.wikipedia.org/wiki/Near_ field_ communication. Accessed 16 October, 2011.

[17] What is NFC (2011) Retrieved from http://www.nfc-forum.org/aboutnfc/. Accessed 16 October, 2011.

[18] Xin Y, Ting Y (2009) A Security Architecture Based on User Authentication of Bluetooth. International forum on Information Technology and Applications, pp. 627 – 629.

[19] Rowe M, Hurman T (2011) Bluetooth Security: Issues, threats and consequences. Retrieved from http://www.pentest.co.uk/documents/wbf_slides.pdf. Accessed 19 June, 2011.

[20] Bluetooth Security Mechanisms (2011) Seguridad Mobile. Retrieved from http://www.seg uridadmobile.com/bluetooth/bluetooth-security/security-mechanisms.html. Accessed 19 June, 2011.

[21] Levy O, Wool A (2005) A Uniform Framework for Cryptanalysis of the Bluetooth E_0 Cipher. 1st International Conference on Security and Privacy for Emerging Areas in Communication Networks (SecureComm), pp. 365 – 373.

[22] Shaked Y, Wool A (2006) Cryptanalysis of the Bluetooth E_0 Cipher using OBDD's. 9th Information Security Conference, LNCS 4176: 187 – 202.

[23] Dunning J (2010) Taming the Blue Beast: A survey of Bluetooth Based Threats. IEEE Security and Privacy, pp. 20 – 27.

[24] Laurie A, Holtmann M, HeMrfurt M (2011) Bluetooth Hacking: The State of the Art. Retrieved from http://trifinite.org/Downloads/trifinite.presentation_22c3_berlin.pdf. Accessed 10 June, 2011.

[25] Becker A (2011) Bluetooth Security & Hacks. Retrieved from http://gsyc.es/~anto/ubicuos2/bluetooth_security_and_hacks.pdf. Accessed 19 June, 2011.

Chapter 6
Security in Vehicular Ad Hoc Networks (VANETs)

Weidong Yang[1]

6.1 Introduction

6.1.1 Overview

As an important component of the intelligent transportation system (ITS) and a novel form of mobile ad hoc network, Vehicular Ad Hoc Networks (VANETs) have attracted much attention from government, academic institutions and industry. In the U.S., the Federal Communications Commission (FCC) has allocated 75 MHz (5.85-5.925 GHz) in the 5.9 GHz band as a new Dedicated Short Range Communications (DSRC) spectrum for vehicular communication. In Europe, the European Telecommunications Standards Institute (ETSI) has also allocated a radio spectrum of 30 MHz (5.875-5.905 GHz) at 5.9 GHz. Similar bands exist in Japan. IEEE has also formed the new IEEE 802.11p task group[1], which focuses on DSRC PHY and MAC layer standard for Wireless Access for the Vehicular Environment (WAVE). Based on the IEEE 802.11p, a higher layer standard IEEE 1609 has been released for trial use[2]. Besides such efforts, many national and international projects devoted to VANETs, such as, the Research and Innovative Technology Administration (RITA) in the United States, the Car-to-Car Communication Consortium (C2C-CC) in European, and the Advanced Safety Vehicle Program (ASV) in Japan.

As is shown in Fig. 6.1, A VANET is a distributed, self-organizing communication network built up by moving vehicles, which contain both Inter-Vehicle (V2V) communications between vehicles and Vehicle-to-Roadside (V2R) communications between vehicles and roadside units (RSUs)[3]. The applications of VANETs can be divided into two major categories: safety

1 Henan University of Technology, China.

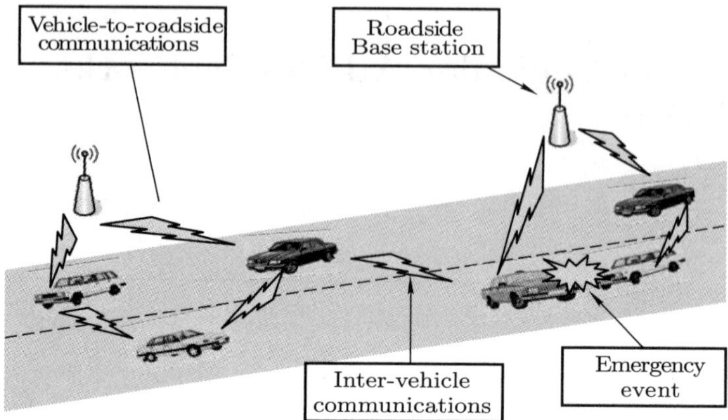

Fig. 6.1 An illustration of VANETs.

and non-safety. Safety applications include collision and other safety warnings, which can be further categorized as safety-critical and safety-related applications. Non-safety applications include real-time traffic congestion and routing information, high-speed tolling, mobile infotainment, and many others.

Despite the fact that vehicles are organized mostly in an ad hoc manner, VANETs have significantly different characteristics compared to traditional mobile ad hoc networks (MANETs). The characteristics of VANETs are given as follows[4,5].

- *Applications.* While most MANET articles do not address specific applications, the common assumption in MANET literature is that MANET applications are identical (or similar) to those enabled by the Internet. In contrast, as we showed above, VANETs have completely different applications.
- *Energy Efficiency.* While in MANETs a significant body of literature is concerned with power-efficient protocols, VANETs enjoy a practically unlimited power supply.
- *Addressing.* Faithful to the Internet model, MANET applications require point-to-point (unicast) with fixed addressing; that is, the recipient of a message is another node in the network specified by its IP address. However, VANET applications often require dissemination of the messages to many nodes (multicast) that satisfy some geo-graphical constraints and possibly other criteria (e.g., directions of movement).
- *Mobility Model.* In MANETs, the random waypoint (RWP)[6] is (by far) the most commonly employed mobility model. However, most existing literature recognized that RWP would be a very poor approximation of real vehicular mobility. When designing a simulation environment, proper vehicular mobility models must be defined in order to produce realistic mobility patterns.

- *Frequent link disconnections.* Unlike nodes in MANETs, vehicles generally travel at much higher speeds, especially on highways. Ascribed to high mobility of vehicles, the topology of a VANET changes rapidly from time to time, causing intermittent communication links.
- *Availability of location information.* Satellite navigation systems are becoming more prevalent in vehicular transportation these days. Making good use of location information by GPS in communication service provision not only can reduce delivery latency of message dissemination (i.e., for road safety services) but also can increase system throughput (i.e., for infotainment services).

The special behavior and characteristics of VANETs create some challenges for vehicular communication, which can greatly impact the future deployment of these networks. A number of technical challenges need to be resolved in order to deploy vehicular networks and to provide useful applications, especially in the aspects of security and privacy[7]. A VANET inherits all the known and unknown security weaknesses associated with MANETs, and could be subject to many security and privacy threats. It is obvious that any malicious behavior of users, such as a modification and replay attack with respect to the disseminated messages, could be fatal to the other users. In addition, the issues in VANET security become more challenging due to the unique features of networks, such as the high mobility of the nodes and the large scale of the network. Furthermore, privacy protection must be achieved in the sense that the user related privacy information, including the driver's name, license plate, speed, position, and traveling routes along with their relationships, has to be protected; while the authorities should be able to reveal the identities of message senders in case of dispute such as a crime/car accident scene investigation, which can be used to look for witnesses. Therefore, it is critical to develop a suite of elaborate and carefully designed security mechanisms for achieving security and privacy preservation in a VANET.

An overview of VANET security can be found in reference [8]. Various consortia are presently addressing VANET security and privacy issues, including the Crash Avoidance Metrics Partnership (CAMP) Vehicle Safety Communications-Applications project, the Vehicle Infrastructure Integration (VII) project, the SeVeCom project, the Embedded Security for Cars (ESCAR) Conference and others. The trial-use standard IEEE 1609.2 (previously named P1556) also addresses security services for VANETs.

6.1.2 VANET Security and Privacy Requirements

The security requirements are derived from primary security goals like confidentiality, integrity and availability. From a review of existing literature[7,14], the general security requirements of a VANET can be derived as authentication, integrity and consistency, confidentiality, availability, access control, non-repudiation and privacy. A security system for safety messaging in a VANET should satisfy the following requirements.

1. *Authentication*

Authentication is a major requirement in VANET as it ensures that the messages are sent by the actual nodes and hence attacks done by the greedy drivers or the other adversaries can be reduced to a greater extent. Authentication in the VANET can be divided into two categories: ID authentication and entity authentication. ID authentication ensures that a message is trustable by correctly identifying the sender of the message. With ID authentication, the receiver is able to verify a unique ID of the sender. The ID could be the license plate or chassis number of the vehicle. Vehicle reactions to events should be based on legitimate messages (i.e., generated by legitimate senders). Therefore, we need to authenticate the senders of these messages. Entity authentication ensures that the recently received message is fresh and live. It ascertains that a message is sent and received in a reasonably small time frame.

2. *Integrity and Consistency*

Integrity requirements demand that the information from the sender to the receiver must not be altered or dropped. The legitimacy of messages also encompasses their consistency with similar ones (those generated in close space and time), because the sender can be legitimate while the messages contains false data.

3. *Confidentiality*

Confidentiality requires that the information flowing from sender to receiver should not be eavesdropped. Only the sender and the receiver should have access to the contents of the message, e.g. instant messaging between vehicles.

4. *Availability*

In safety applications like post-crash warning, the wireless channel has to be available so that approaching vehicles can still receive the warning messages. If the radio channel goes out (e.g. jamming by an attacker), then the warning cannot be broadcasted and the application itself becomes useless. Hence availability should be also supported by alternative means.

5. *Access Control*

Access control is necessary for an application that distinguishes between different accessing levels of a node or infrastructure component. This is established through specific system-wide policies, which specifies what each node is allowed to do in the network. For instance, an authorized garage may be allowed to fully access wireless diagnostics, whereas other parties may only be granted limited accesses. Another form of access control can be the exclusion of misbehaving nodes (e.g. by an intrusion detection system using a trust management scheme) from the VANET by certificate revocation or other means.

6. *Non-repudiation*

Drivers causing accidents should be reliably identified. A sender should not be able to deny the transmission of a message (it may be crucial for investigation to determine the correct sequence and content of messages exchanged before the accident).

7. *Privacy*

Privacy is an important factor for the public acceptance and successful deployment of VANETs. With vehicular networks deployed, the collection of vehicle-specific information from overheard vehicular communications will become particularly easy. Then inferences on the drivers' personal data could be made, and thus violate her or his privacy. The vulnerability lies in the periodic and frequent vehicular network traffic messages which will include, by default, information (e.g., time, location, vehicle identifier, technical description, trip details) that could precisely identify the originating node (vehicle) as well as the drivers' actions and preferences. Hence, the privacy of drivers against unauthorized observers should be guaranteed.

6.1.3 Security Threats in Vehicular Ad Hoc Networks

A VANET can be compromised by an attacker from manipulating either vehicular system or the security protocols. Hence two kinds of attacks can be visualized against vehicular systems: attacks against messages and attacks against vehicles. Next, we explore the most significant vulnerabilities of vehicular communications.

1. *In the case of an accident*

In the worst case, colluding attackers can clone each other, but this would require retrieving the security material and having full trust between the attackers. In cases where liability is involved, drivers may be tempted to cheat with some information that can determine the location of their car at a given time.

2. *In-transit traffic tampering*

Any node acting as a relay can disrupt communications of other nodes: it can drop or corrupt messages, or meaningfully modify messages, so that the reception of valuable or even critical traffic notifications or safety messages can be manipulated. Moreover, attackers can replay messages (e.g., to illegitimately obtain services such as traversing a toll check point). In fact, tampering with in-transit messages may be simpler and more powerful than forgery attacks.

3. *Masquerading*

The attacker actively pretends (impersonates) to be another vehicle by using false identities and can be motivated by malicious or rational objectives. Message fabrication, alteration, and replay can also be used towards masquerading. A masquerader can be a threat: consider, for example, an attacker masquerading as an emergency vehicle to mislead other vehicles to slow down and yield.

4. *Privacy violation*

With vehicular networks deployed, the collection of vehicle specific information from overheard vehicular communications will become particularly easy. Then inferences on the drivers' personal data could be made, and thus violate her or his privacy. The vulnerability lies in the periodic and frequent vehicular network traffic. In all such occasions, messages will include, by default, information (e.g., time, location, vehicle identifier, technical description, trip details) that could precisely identify the originating node (vehicle) as well as the drivers' actions and preferences.

5. *Denial of Service (DoS)*

The attacker may want to bring down the VANET or even cause an accident. There are many ways to perform this attack, either by sending messages that would lead to improper results or by jamming the wireless channel (this is called a Denial of Service, or DoS attack) so that vehicles cannot exchange safety messages.

6. *Hidden vehicle*

In this scenario, a vehicle broadcasting warnings will listen for feedback from its neighbors and stop its broadcasts if it realizes that at least one of these neighbors is better positioned for warning other vehicles. This reduces congestion on the wireless channel. A hidden vehicle attack consists in deceiving vehicle A into believing that the attacker is better placed for forwarding the warning message, thus leading to silencing A and making it hidden (has stopped broadcasting).

7. *Tunnel*

Since GPS signals disappear in tunnels, an attacker may exploit this temporary loss of positioning information to inject false data once the vehicle leaves the tunnel and before it receives an authentic position update. The physical tunnel in this example can also be replaced by an area jammer from the attacker, which results in the same effects.

8. *Sinkhole attack*

In sinkhole attack, an intruder attracts surrounding nodes with unfaithful routing information, and then performs selective forwarding or alters the data passing through it. The attacking node tries to offer a very attractive link e.g. to a gateway. Therefore, a lot of traffic bypasses this node. Besides simple traffic analysis, other attacks like selective forwarding or denial of service that can be combined with the sinkhole attack.

9. *Wormhole attack*

The attacker connects two distant parts of the ad hoc network using an extra communication channel as a tunnel. As a result, two distant nodes assume they are neighbors and send data using the tunnel. The attacker has the possibility of conducting a traffic analysis or selective forwarding attack. This also extends the range of the attacker.

10. *Sybil attack*

Large-scale peer-to-peer systems face security threats from faulty or hostile remote computing elements. To resist these threats, many such systems employ redundancy. However, if a single faulty entity can present multiple identities, it can control a substantial fraction of the system, thereby undermining this redundancy. The Sybil attack especially aims distributed system environments. The attacker tries to act as several different identities/nodes rather than one. This allows him to forge the result of a voting used for threshold security methods.

11. *On-board tampering*

Other than communication protocols, an attacker may select to tinker with data (e.g., velocity, location, status of vehicle parts) at their source, tampering with the on-board sensing and other hardware. In fact, it may be simpler to replace or by-pass the real-time clock or the wiring of a sensor, rather than modifying the binary code implementation of the data collection and communication protocols.

6.2 Security Architecture Framework for Vehicular Ad Hoc Networks

6.2.1 Overview

Currently, many of the researches in VANET are paying more attention on the development of a proper MAC layer (the definition of the MAC and physical layer protocols has greatly progressed, for instance IEEE 802.11p, a specially designed version of IEEE 802.11) rather than security architecture and protocols for VANET. The most prominent industrial effort in this domain is carried out by Car 2 Car Communication Consortium, the IEEE 1609.2 working group, the NoW project and the SeVeCom project with all of them developing VANET Security architecture. All of them take the use of Certification Authority (CA) and public key cryptography to protect V2V and V2I messages as their basic elements. It has now become an established consensus that public key cryptography is the way to go about for VANETs. This is mainly due to the fact that the messages are broadcasted and one-to-one communication is not the norm. Due to this fact, symmetric key cryptography will incur huge costs in frequent key establishment procedures and they are also difficult to implement as the nodes are constantly on the move. For all the perspective security protocols, message authentication, integrity and non-repudiation, as well as protection of private user information are identified as primary requirements.

6.2.2 PKI for Vehicular Ad Hoc Networks

Vehicles are registered in different states and they're huge in numbers. They will travel long distances so that they can be well beyond their registration areas. All of these are requiring a robust and flexible key management scheme. The involvement of authorities in vehicle registration implies the need for a certain level of centralization. Vehicles not only have to be identified by base stations, but also have to be identified by each other (without invoking any server), so that communications by base station (as in cellular networks) is not enough for VC, and this creates a problem of scalability. In addition, symmetric cryptography does not provide the non-repudiation property that allows the accountability of drivers' actions (e.g., in the case of accident reconstruction or finding the originator of forgery attacks). Hence, the use of public key cryptography is a more suitable option for deploying vehicular communications security.

This implies the need for a public key infrastructure (PKI). As stated above, VANET is usually a hybrid network with the possibility to access a stationary network at least temporarily, so that using a centralized PKI approach with a TTP which issues certificates and revokes them is an appropriate idea. Therefore, a PKI with the certification authority CA (the trust center) is used to introduce trust within the network. Fig. 6.2 shows the basic setup of the PKI. In order to communicate, a node (CA) has to be registered at the trust center. By fulfilling the registration process, the vehicles can get a certificate signed with the key of the CA. The CA is responsible for checking if the right vehicle get the right key and if the vehicle is worthy of trusting before issuing the signed certificate. Every subscriber within the network knows the public key of the CA and can check the validity of any public key certificate issued by the CA. Therefore, any two vehicles can exchange and validate their public keys without having access to any other node or gateway. If the certificates are valid, the vehicles can trust each other and establish a secure connection.

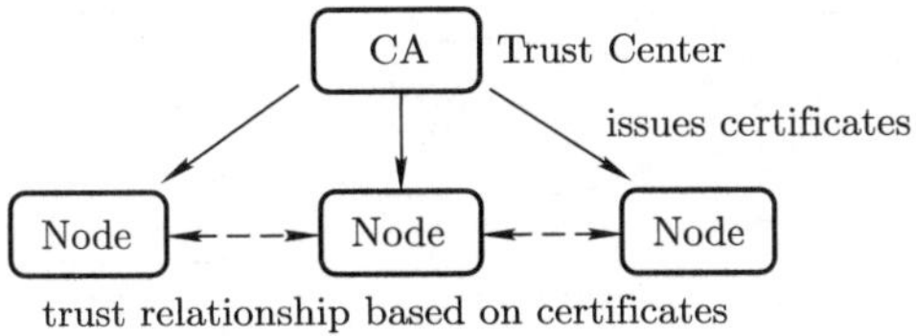

Fig. 6.2 Public key infrastructure.

6.2.3 Trusted Architecture for Vehicular Ad Hoc Networks

In this subsection, we first discuss IEEE 1609.2 standard which specifies methods of securing Wireless Access in Vehicular Environment (WAVE) messages against various attacks. We then describe security architecture for VANETs based on the PKI and security hardware[9]. A secure VANET communication scheme based on TPMs[10] is also given.

6.2.3.1 IEEE 1609.2 Security Framework

The IEEE 1609 communication standards, also known as Dedicated Short Range Communications (DSRC) protocols, have emerged recently to enhance 802.11 to support wireless communications among vehicles for the roadside infrastructure. The IEEE 1609.2 standard addresses the issues of securing WAVE messages, in order to fight against eavesdropping, spoofing, and other attacks. The components of the IEEE 1609.2 security infrastructure which are based on industry standards for public key cryptography, includes support for elliptic curve cryptography (ECC), WAVE certificate formats, and hybrid encryption methods, in order to provide secure services for WAVE communications that are shown in Fig. 6.3. To support core security functions such as certificate revocation, the security infrastructure is also needed to be responsible for the administrative functional necessities. Note that certificate revocation is essential to any security system based on the public key infrastructure, which has not been addressed in the current IEEE 1609.2 by considering the unique features of vehicular networks. In addition, IEEE 1609.2 does not define driver identification and privacy protection, and has left a lot of issues open.

6.2.3.2 Security architecture based on security hardware and the PKI

Here security hardware means two hardware modules among the vehicle onboard equipment for security, namely the event data recorder (EDR) and the tamper-proof device (TPD). Whereas the EDR only provides tamper-proof storage, the TPD also possesses cryptographic processing capabilities. The EDR has the function of recording the vehicle's critical data, such as position, speed, time, etc., during emergency events, similar to an airplane's black box.

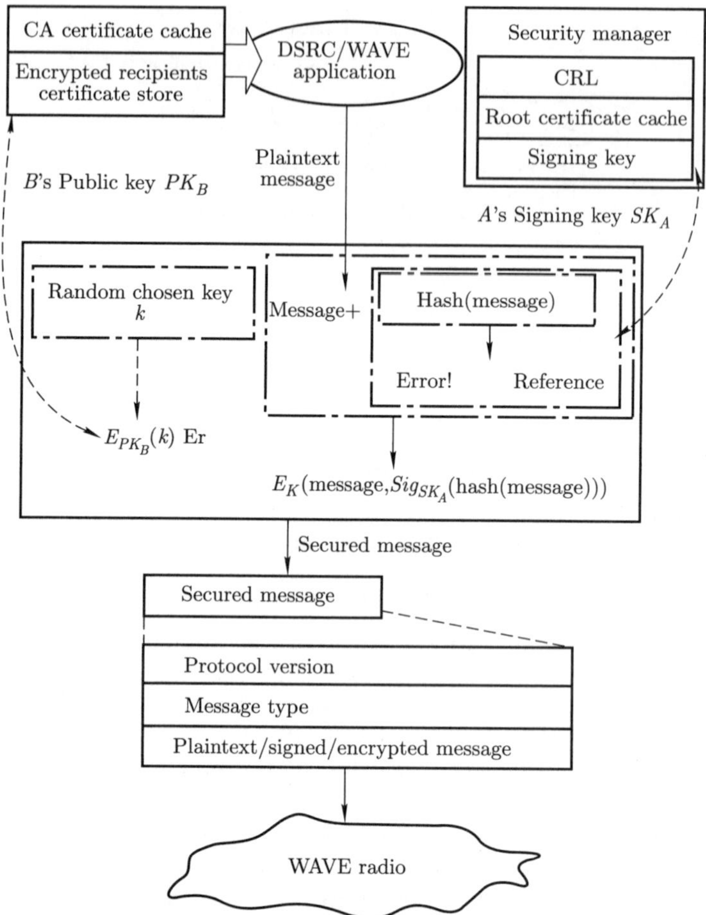

Fig. 6.3 The IEEE 1609.2 security services framework for creating and exchanging WAVE messages between WAVE devices.

These data are useful in accident reconstruction and the attribution of liability. EDRs have already been installed in a lot of road vehicles, especially trucks. These can also record the safety messages received if critical events happen.

An owner or a mechanist can easily accesse the vehicle electronics, especially the data bus system. Therefore, the cryptographic keys of a vehicle need proper hardware protection, namely a TPD. The TPD will take care of storing all the cryptographic material and performing cryptographic operations, especially signing and verifying safety messages. After connecting a set of cryptographic keys to a given vehicle, the TDP guarantees the accountability property as long as it remains inside the vehicle. The TPD needs to be independent from its external environment. It should own its own clock and have a battery that is periodically recharged from the vehicle's electric circuits. The general secure architecture based on security hardware and PKI is given by Fig. 6.4.

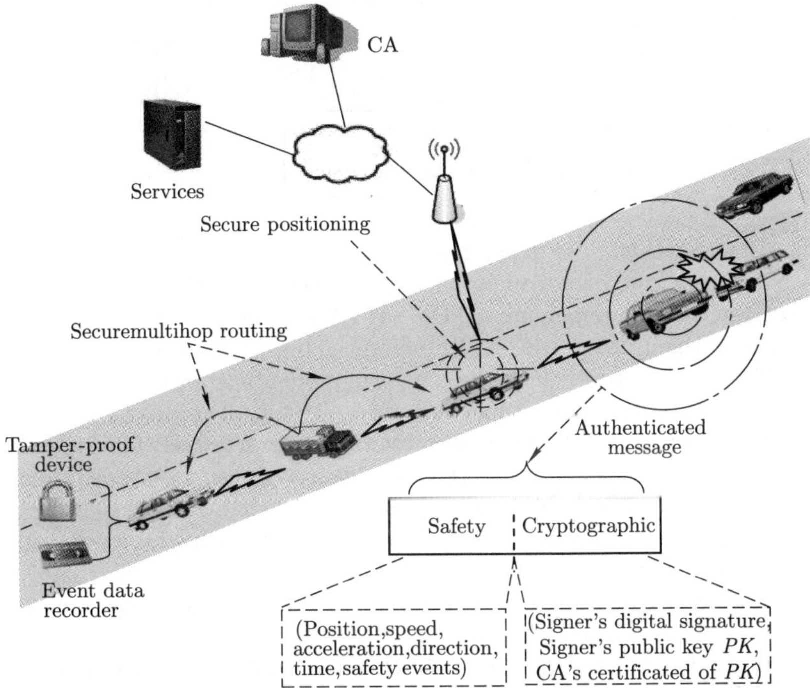

Fig. 6.4 Security architecture based on security hardware and the PKI.

6.2.3.3 Secure VANET communication scheme based on TPMs

The Trusted Platform Module (TPM)[11] can be integrated into the vehicle onboard equipment for implementing the security requirements. It is a general purpose of hardware chip designed for secure computing. A TPM is a piece of hardware, requiring a software infrastructure, which is able to protect and store data in shielded locations. A TPM has also cryptographic capabilities such as a SHA-1 engine, an RSA engine, and a random number generator. Fig. 6.5 illustrates the main components of a TPM.

Here are the two levels where the security model are using TPMs to secure VANET works. The first level permits a trusted channel to be established between any two vehicles. This means that the two vehicles are satisfied that each is running an untampered version of the security software, and that no intentional data attack or Sybil attack is being attempted. The second level aims at information verification. It builds on trusted channels, and is to ensure that a vehicle's configuration does not contain erroneous readings.

Implementing trusted channels relies directly on the TPM's attestation

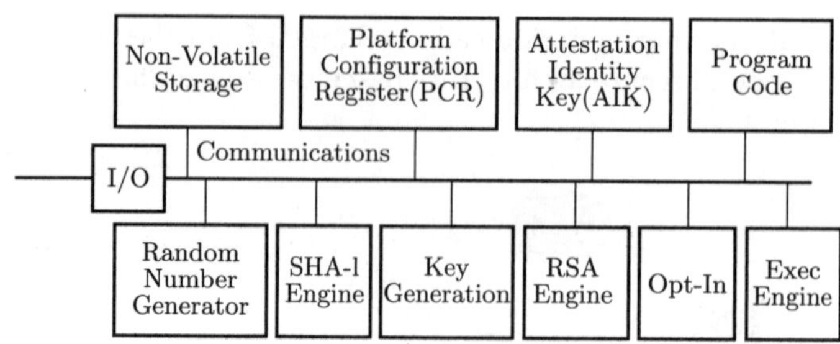

Fig. 6.5 Architecture of a TPM.

mechanism. A vehicle can trust another if the latter can demonstrate that its software has not been tampered and the source of the software can be verified. The issue in deploying a TPM on VANET nodes is to assign roles to the actors in the TPM protocols. In reference [10] the following are assumed:

(1) Car manufacturers sign the platform credentials for their vehicles. To assume that a manufacturer takes responsibility for all embedded devices on their vehicles is rational. Further, manufacturers are relatively few in numbers and are well-known in the sense that certificates signed by these principals should be recognizable to all vehicles and automobile authorities.

(2) Automobile authorities are responsible for organizing technical reviews. In most countries, car owners are obliged to submit their cars to a technical review every 2 to 3 years. If a car fails the technical review, it cannot be driven on the road. Automobile authorities are thus well-known principals that can act as privacy CAs that can sign AiK credentials.

The TPM gives us a means to securely attribute a vehicle identifier. This can be signed by an automobile authority. When vehicles exchange messages, we can use the attestation protocol and then ensure the integrity and authenticity of these messages.

The second level of security, which is information verification, is based on three simple procedures.

(1) Auto-measuring. A vehicle's software maintains data on the vehicle's acceleration and deceleration capabilities, as well as related data such as tire denseness (which embedded devices are now able to measure). These values evolve so the vehicle continuously updates them. These values are obviously important for the platoon scenario where neighboring vehicles need to agree on minimal distances.

(2) Challenge-response protocol. This procedure is needed to find out unintentional errors in information transmitted by a vehicle that are due to permanent errors in the sensor of the vehicle. Vehicles that are close together should possess the same readings for many information types, for example, such as temperature, time, and location. It aims to permit a vehicle to challenge another with respect to any of these readings.

(3) Technical review. The automobile authorities organize technical reviews. The vehicles with VANET functionality must include reviews of the

correct functions of all sensor devices. Further, it is expected that any changes that need to be made to the application software are made at this moment. Since the TPM can only be used to help verify that the software on a platform has not been tampered with, it is very important to know that the absence of security flaws or bugs in the software itself does not guarantee. The three procedures conduce to detect and isolate permanent errors in readings.

Figure 6.6 shows the different components of the embedded architecture and the data flow. As it shows, for instance, in auto-measuring, sensors embedded in vehicle give results of their measures to the application. Then the application asks the TPM to sign the data. The TPM checks the PCR value associated with this application and signs data provided by the application. Then it will store this data in a dedicated repository.

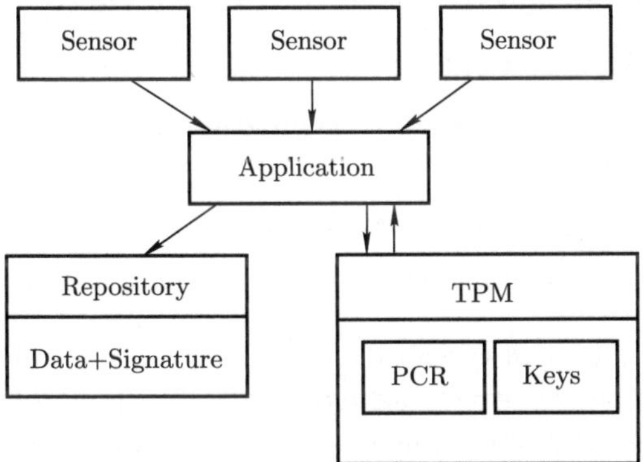

Fig. 6.6 The embedded architecture.

In order to detect unintentional errors, the details for challenging another vehicle are given in Fig. 6.7. The challenger sends a query about data it

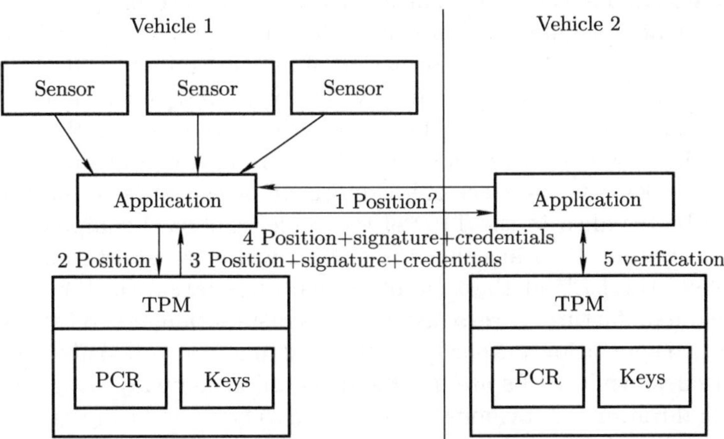

Fig. 6.7 The challenge-response protocol.

can verify, the current position in the example. Then the challenged vehicle collects the appropriate data, and gives this data to its TPM. The TPM checks the PCR values associated with this application and signs data. The application sends to the challenger the signed data and associated credential. The challenger verifies the signature and compares the given position to its own current position to detect misconfiguration of the positioning unit of the challenged vehicle.

6.2.4 Key Management and Authentication Scheme

In this section we present the scheme of key management and authentication under the security architecture based on the PKI and security hardware.

6.2.4.1 Key Management

We will address below the issues of cryptographic key distribution, certification, and revocation.

1. *Cryptographic Information Types and Key Distribution*

To be part of a VANET, each vehicle has to store the following cryptographic information[12]:

(1) an electronic identity called an electronic license plate (ELP) issued by a government, alternatively an electronic chassis number (ECN) issued by the vehicle manufacturer. These identities (further referred to simply by ELP) should be unique and cryptographically verifiable (this can be achieved by attaching a certificate issued by the CA to the identity) in order to identify vehicles to the police in case this is required (identities are hidden from the police). It seems to the physical license plates, the ELP should be changed when the owner changes or moves, e.g., to a different region or country.

(2) Anonymous key pairs that are used to preserve privacy. An anonymous key pair is a public/private key pair that is authenticated by the CA. However it contains neither information about nor public relationship with (i.e., this relationship cannot be discovered by an observer without a special authorization) the actual identity of the vehicle (i.e., its ELP). Usually, a vehicle will possess a set of anonymous keys to prevent tracking.

Now the ELP is the electronic equivalent of the physical license plate, it should be installed in the TPD of the vehicle onboard equipment using a similar procedure. It means that the governmental transportation authority will preload the ELP at the time of vehicle registration (in the case of the ECN, the manufacturer is responsible for its installation at production time).

The transportation authority or the manufacturer preloads anonymous keys. Besides, while ELPs are fixed and should accompany with the vehicle for a long duration, anonymous key sets have to be periodically renewed after all the keys have been used or their lifetimes have expired. During the periodic vehicle checkup (typically yearly) or by similar procedures this renewal can be done.

Over and above, the ELP and anonymous keys, each vehicle should be preloaded with the CA's public key.

2. *Key certification*

CA will be responsible for issuing key certificates to vehicles. Here are two solutions.

(1) Governmental transportation authorities: The corresponding transportation authorities (which are usually regional) will register vehicles in different countries. The advantage of this option is that the certification procedure will be under the direct control of the concerned authority. Although the ELP and keys of each vehicle are certified by a regional authority in a given country, vehicles from different regions or countries should be able to authenticate each other. This problem is usually solved by including the certificate chain leading to a common authority, but in the case of VANET, it would tremendously increase the message overhead. This certificate chain can be replaced by a single certificate by making the CA of the traveling vehicle's transit. Also can destination region recertify the ELP and the anonymous keys of the vehicle after verifying them with the public key of the CA that registered the vehicle? This requires the installation of base stations at the region borders.

(2) Vehicle manufacturers: Considering the limited number and the trust already endowed in them, certificates can also be issued by vehicle manufacturers. The advantage of this approach is to reduce overhead. In fact, in order to be able to verify any other vehicle it encounters, which is not the case if the CA is a local authority, each vehicle will need to store a small number of manufacturer public keys. However, this approach could lead to non-governmental institutions being involved in law enforcement mechanisms.

3. *Key revocation*

The owner's identity, certified and issued by a CA, is connected with the public key by a public key certificate. Various attacks including man-in-the-middle attacks and impersonation attacks can be effectively prevented with the help of a public key certificate. However, a user's certificate could be repealed due to some unexpected reasons. For instance, to maintain system security, the certificate should be repealed once the private key corresponding to the public key specified in the certificate is identified as compromised.

The traditional PKI architecture use certificate revocation scheme most through the certificate revocation list (CRL), a list of revoked certificates stored in central repositories prepared in CAs. Based on such centralized architecture, alternative solutions to CRL could be a certificate revocation system (CRS), certificate revocation tree (CRT), the Online Certificate Status Protocol (OCSP)[13], and other methods. Usually, these schemes are required to be highly available of the centralized CAs, where frequent data transmission with vehicles to obtain timely revocation information may cause significant overhead. Therefore, the centralized CRL architecture may only exist in fantasy with the high-speed mobility and large quantity of network entities in VANETs.

In order to solve the problem, Lin et al.[14] came up with a novel RSU-aided certificate revocation (RCR) mechanism for performing certificate revocation. As illustrated in Fig. 6.8, there are three types of network entities: the authority (denoted as CA), RSUs, and vehicles. The relationship between these three is explained as follows. The CA manages the RSUs, and both of them are assumed to be trustworthy. The RSUs are connected to the Internet through either wired Ethernet or WiMAX, or any other networking technology. Furthermore, the CA provides each RSU a secret key, while the corresponding public key is an identity string containing the name of the RSU, the physical location, and the authorized message type. By this approach, the messages can be signed by an RSU with the help of an identity-based signature.

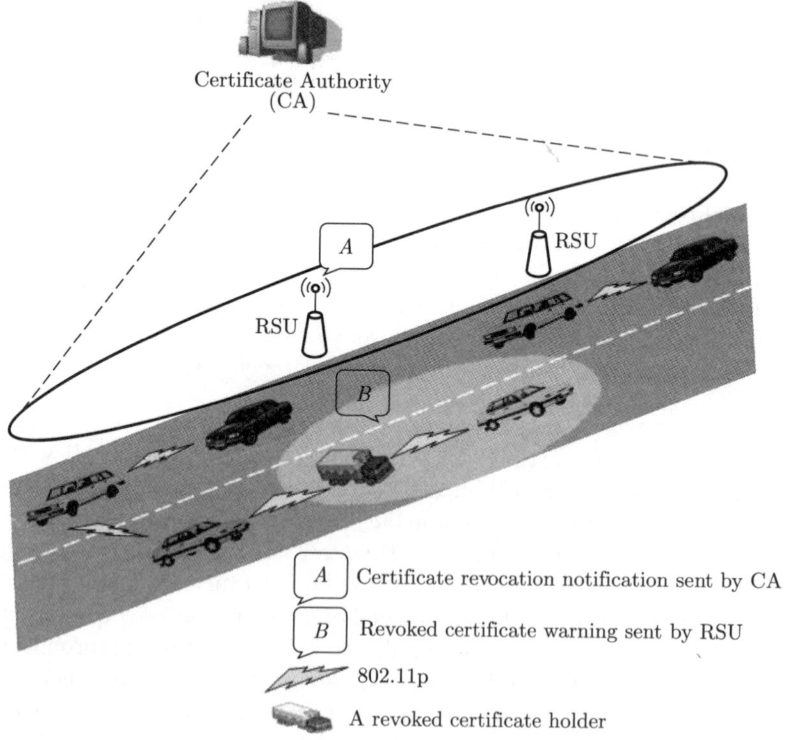

Fig. 6.8 The RSU-aided key revocation scheme.

The CA will inform all the RSUs about a certificate revocation once a certificate is revoked. Each RSU then checks the status of the certificates contained in all the messages broadcasted by the passing vehicles. If a certificate has been confirmed as revoked, the RSU will broadcast a warning message such that all other approaching vehicles can update their CRLs and avoid communicating with the compromised vehicle. Since vehicle movement can be predicted based on its driving conditions (e.g., direction, speed, position), the RSU can further notify all neighboring RSUs of where the compromised vehicle may go. In addition, due to RSUs' normally sparse location, a rather limited number of vehicles will be notified even if all the RSUs broadcast the corresponding message. Therefore, in order to make the warning message disseminate more effective, the warning message among vehicles can be forwarded through inter-vehicle communications, that is to say, disseminated by each vehicle, hop by hop, throughout its predefined lifetime.

However, in order to avoid being detected, while passing through an RSU, a compromised vehicle may intentionally disable message broadcasting. This is also referred as a silent attack, which can easily be handled by granting every RSU the privilege of signing the certificate of each vehicle. In this case, whenever a vehicle passes through an RSU, the vehicle asks the RSU to sign its certificate, where the signature serves as evidence that can demonstrate its authenticity and legitimacy to other vehicles. The corresponding messages will be ignored if a vehicle is using a certificate that has not been verified by an RSU for a certain period of time and is discovered by a neighbor vehicle. Therefore, according to resisting compromised vehicles, the VANET can gain the security and safety with the least amount of effort.

6.2.4.2 Authentication Scheme

The safety messages in VANETs can be classified into three classes[12], based on their properties related to privacy and real-time constraints, as shown in Table 6.1. Traffic information messages are used to disseminate traffic conditions in a given region and thus affect public safety only indirectly (by preventing potential accidents due to congestion); hence they are not time-critical. General safety-related messages are used by public safety applications such as cooperative driving and collision avoidance and hence should satisfy stringent constraints such as an upper bound on the delivery delay. Liability-related messages are distinguished from the previous class because they are exchanged in liability-related situations such as accidents. Therefore, the liability of the message originator should be determined by revealing his identity to the law enforcement authorities. A common property of all the message classes is that they are mainly standalone and there is no content dependency among them. Apart from data specific to traffic events, position, speed, direction, and acceleration of the vehicle are also concluded within a typical safety message. In case the sender is trapped in an abnormal situation, for instance, an accident, these data would help receivers compute their positions concerning the sender and examine if they are in danger.

All the message classes share another common property, in which they don't contain any sensitive information, and confidentiality is not required. As a result, the exchange of safety messages in a VANET needs authentication but not encryption. For message authentication, there is a simplest and most efficient way, which is, assigning to each vehicle a set of public/private key pairs, that will allow the vehicle to digitally sign messages and thus authenticate itself to receivers.

Table 6.1 Message classes and properties

Class/Property	Legitimacy	Privacy Protection		Real-time Constraints
		Against Others	Against Police	
Traffic Information	yes	yes	yes	
General Safety Messages	yes	yes	yes	yes
Liability-Related Messages	yes	yes		yes

A practical authentication scheme is shown as follows. Before a vehicle sends a safety message, it signs it with its private key and includes the CA's certificate as follows:

$$V \to * : M, sigSK_V[M|T], Cert_V$$

where V designates the sending vehicle, $*$represents all the message receivers, M is the message which is actually hashed before being signed, SK_V is V 's private key, $|$ is the concatenation operator, and T is the timestamp to ensure message freshness (it can be obtained from the security device TPD). It should be noted that, because of the burden of the inherent preliminary handshake where the communicating parties exchange the nonces, using nonces instead of timestamps is not desirable. Using sequence numbers also incurs overhead as they need to be maintained. $Cert_V$ is the public key certificate of V later.

Using the certificate, the receivers of the message have to extract and verify the public key of V, and then verify V 's signature using its certified public key. In order to do this, the receiver should have the public key of the CA, which can be preloaded as described above. If the message is sent in an emergency context, which means that it belongs to the liability-related class, this message should be stored (including the signature and the certificate) in the EDR for further potential investigations in the emergency.

This authentication scheme has failed in taking the scalability issue and resulted communication overhead into consideration. Furthermore attaching a digital signature and a certificate to each safety message for the sake of security inevitably creates overhead that can be larger than the message itself. Therefore Zhang et al.[15] proposes an RSU-aided message authentication scheme, called RAISE, which explores the unique features of VANETs by employing RSUs to assist vehicles in authenticating messages. With RAISE, when an RSU is detected nearby, vehicles start to associate with the RSU. Then, the RSU assigns a unique shared symmetric secret key and a pseudo ID that is shared with other vehicles. With the symmetric key, each vehicle generates a symmetric keyed-hash message authentication (HMAC) code, and then broadcasts a message by signing the message with the symmetric HMAC code instead of a PKI-based message signature. Other vehicles receiving the messages signed with the HMAC code are able to verify the message by using the notice about the authenticity of the message disseminated by the RSU.

The detailed implementation of RAISE is presented in the following. The notations are listed in Table 6.2 for ease of presentation.

Table 6.2 Notations

Notations	Descriptions
R_i:	the i-th RSU
V_i:	the i-th vehicle
M_i:	the message sent by V_i
K_i:	the key shared between V_i and R_i
ID_i:	a pseudo identity of V_i assigned by R
U:	an entity, which could be an RSU R or a vehicle V_i
T:	the current time
PK_U:	the public key of U
SK_U:	the private key of U
C_U:	U's certificate
$\{m\}_{SK_U}$:	U's digital signature on m
$H(\cdot)$:	a one-way hash function such that SHA-1
$HMAC(\cdot)$:	a keyed-hash message authentication code
$\|\|$:	message concatenation operation

1. *Symmetric key establishment*

Once a vehicle V_i detects that there is an RSU R_i nearby, V_i initiates a mutual authentication process and establishes a shared secret key with R_i. This can be achieved by adopting the Diffie-Hellman key agreement protocol secured with public key based signature scheme. The mutual authentication and key agreement processes are shown as follows:

$$V_i \rightarrow R : g^a, \{g^a\}_{SK_{V_i}}, C_{V_i}$$
$$R \rightarrow V_i : ID_i\|g^b, \{ID_i\|g^a\|g^b\}_{SK_R}, C_R$$
$$V_i \rightarrow R : \{g^b\}_{SK_{V_i}}$$

where g^a and g^b are elements of the Diffie-Hellman key agreement protocol, and the shared key between R_i and V_i is $K_i \leftarrow g^{ab}$. When receiving the first message from V_i, R_i can verify V_i's public key PK_{V_i}, and then use PK_{V_i} to verify the signature $\{g^a\}_{SK_{V_i}}$ on g^a. In a similar manner, V_i authenticates R_i. If the above three flows succeed, the mutual authentication process is done. At the same time, in the second flow, R_i assigns a pseudo identity ID_i to the vehicle V_i. The pseudo ID is uniquely linked with K_i. With ID_i, R_i can know which vehicle sends the message, and can further verify the authenticity of the message with their shared symmetric key. Therefore, R_i maintains an ID-Key table in its local database.

2. *Hash aggregation*

Once the vehicle V_i obtains the symmetric key K_i from the RSU R_i, V_i uses K_i to compute the message authentication code $HMAC(ID_i||M_i)$ on $ID_i||M_i$, where ID_i is V_i's pseudo identity assigned by R_i and M_i is the message to be sent. Then, V_i one-hop broadcasts $ID_i||M_i||HMAC(ID_i||M_i)$. Since K_i is only known by R_i in addition to V_i itself, only R_i can verify M_i. Thus, to make other vehicles be able to verify the authenticity of M_i, and at the same time to reduce communication overhead, the RSU R_i is responsible to aggregate multiple authenticated messages in a single packet and to send it out. The detailed process is shown as follows:

(1) R_i checks whether the time interval between the current time and the time when R_i sent the last message authenticity notification packet is less than a predefined threshold. If so, go to Step 2. Otherwise, go to Step 4.

(2) When R_i receives a message, $ID_i||M_i||HMAC(ID_i||M_i)$, sent by the vehicle V_i, R_i first checks whether ID_i is in R_i's ID-Key table. If yes, go to Step 3. Otherwise, go to Step 4.

(3) R_i uses ID_i's K_i to verify $HMAC(ID_i||M_i)$. If it is valid, R_i computes $H(ID_i||M_i)$ and then go to Step 1. Otherwise, drop the packet.

(4) R_i aggregates all hashes generated at Step 3, i.e., $HAggt = H(ID_1||M_1) ||H(ID_2||M_2)||\ldots||H(ID_n||M_n)$, and signs it with its private key SK_{R_i}. Then, R_i one-hop broadcasts $HAggt||\{HAggt\}_{SK_{R_i}}$ to vehicles within its communication range.

3. *Verification*

When the other vehicles sent messages to a vehicle, received vehicles only buffers the received messages in its local database without verifying them immediately. The buffered record has the following format: M_i, ID_i, $H(ID_i||M_i)$. Once vehicles obtain the signed packet $HAggt||\{HAggt\}_{SK_{R_i}}$ from the RSU, they are able to verify the buffered messages one by one. First, vehicles use the RSU's public key PK_{R_i} to verify the signature $\{HAggt\}_{SK_{R_i}}$. If it is valid, vehicles will check the validity of the previously received messages buffered in the record in the local database. This is done by comparing whether there is a match between the buffered record with the de-aggregate message.

A vehicle generates a HMAC for each launched message with RAISE. The HMAC can only be generated by the vehicle that has the key assigned by the RSU. When the adversary tempers a message, the RSU cannot find a responding validation key that can compute a matching HMAC for the message, and therefore the tempered message will be ignored. On the side, for each vehicle, there is a unique key stored in the ID-Key table in the RSU side. When an RSU finds out a key that can verify the HMAC, the RSU knows the identity of the message sender, and as a result the source is authenticated.

6.3 Secure Communication protocols for Vehicular Ad Hoc Network

6.3.1 Overview

There are many communication patterns in the VANET. Different communication patterns require different secure mechanisms to thwart security and privacy infringements. Therefore, we have to identify first which communication protocols will finally be used. In reference [16], the SeVeCom project extrapolates three basic communication patterns:

(1) Beaconing (Periodic, single-hop broadcasts, containing e.g. a vehicle's location, heading etc.).

(2) Restricted Flooding/Geocast (Multi-hop broadcast over a certain number of hops restricted by TTL or by specified geographic destination region).

(3) Geographic uni-cast routing (Multi-hop, hop-by-hop forwarding of packets, either for uni-cast end-to-end connections for any cast requests or for subsequent flooding/geocast in a remote destination region).

Basic questions about secure communication regard to which and how security mechanisms can be used to secure communication protocols, and how these security mechanisms can be integrated with the actual functional components, like the routing or medium access. Therefore, the usage of communication patterns instead of concrete protocols has the advantages that we stay independent of the implementation details and security mechanisms can easily be adapted to similar communication protocols.

6.3.2 Secure Beaconing

In VANETs, beaconing denotes a mechanism which broadcasts information periodically over a single hop, which means that they are not relayed by receiving nodes. Besides some identifiers, the information typically includes the vehicle's own position and additional information like speed or heading direction. Beacons are usually not forwarded, i.e. are consumed after one hop. This kind of communication is useful for instance for all cooperative awareness applications.

As a basic goal, a receiver needs to be able to verify authenticity and integrity of beacons. This means that a vehicle must be able to trust in the content of a beacon message in a way that

- the sender is actually a valid participant of the network (e.g., a vehicle, RSU, traffic sign, etc.),
- the identified sender has sent the message, not another one,
- the data is up-to-date,

- the data has not been altered.

During the setup of the system, the secure beaconing component is hooked into the data delivery path. When we assume network layer beaconing, the secure beaconing component is attached between the network and a link layer. Using this hook, the secure beaconing component will process the beacon data both upon sending and upon reception of a beacon.

When a beacon message is lined up to be sent, the hook redirects the message to the secure beaconing component. To be able to scan the content of the beacon, the message format must be known to the secure beaconing component, at least to some extent.

As mentioned earlier, typical beacons will include at least

- the current vehicle identifier (pseudonym) X,
- the current vehicle location $locX$.

In addition, the secure beaconing also requires a current time stamp (t_c) to be included in the beacon message in order to be able to ensure freshness of beacons.

For both efficiency and security reasons, these fields should not be duplicated in a beacon message. Hence, the implementation has to reuse existing fields. In case that the required fields are not included already, they have to be appended by the secure beaconing. Moreover, even if the required fields are already included, secure beaconing has to ensure that they comply with the security requirements. For instance, if the application has already added the field for the vehicle position, but this position information is not accurate enough for security reasons, another, appropriate location has to be appended by the secure beaconing.

Finally, the PAYLOAD should contain:

$$\text{PAYLOAD} = X|locX|\cdots$$

After these preprocessing steps, the component uses signing capabilities of the identification and trust management module. Moreover, the current time t_c is returned together with the signature, as the hardware security module provides a function to sign with timestamp.

After that, the beacon message will comprise payload, timestamp, signature, and certificate:

$$\text{BEACON} = \text{PAYLOAD}|t_c|sigSK_x(\text{PAYLOAD}|t_c)|certPK_x$$

The signed BEACON will then be returned into the data delivery path.

When a beacon arrives at a vehicle, it is passed over to the secure beaconing component via the hooking interface. The component will first check the attached signature by using the verify method of the identification management module. If the signature can be verified, further post-processing is applied, like the freshness check to prevent replay of old messages. If the signature is invalid, the message is either discarded immediately or marked as invalid by the component. The choice depends on whether applications also want to process invalid packets and should be configurable.

After the signature check, which includes a certificate check, it can be guaranteed that

- The message was sent once by the given sender X,
- The message has not been altered,
- The sender is a valid network participant.

Moreover, as the messages must not be replayed from vehicles passing by earlier, the freshness check needs to validate that the message's timestamp is recent. This requires determining the current time, which is provided by the hardware security module.

Noted that the freshness check should explicitly tolerate propagation delay, an allowed deviation of several seconds seems reasonable to prevent large-scale replay. This treatment also has the advantage that clocks do not need to be tightly synchronized. Nevertheless, if an older message is received, it is discarded.

Due to their high frequency, a number of challenges on the application of crypto mechanisms arise:

(1) Because of their frequency, beacons can cause a substantial part of the overall channel load. This situation is aggravated if every packet has to carry a complete set of security data like signature and certificate. Therefore, it would be desirable to reduce the channel load by more sophisticated security solutions. At the same time, each packet should be self-contained, i.e. authentication and integrity checks should be achievable without the context of other packets to allow for fast evaluation of time-critical packets.

(2) A similar problem due to high frequency of beacons originates from the computational requirements of asymmetric crypto operations. It is well known that creation and verification of asymmetric signatures can consume considerable amount of time. For example, if we assume beacons to be sent with frequency f and the current vehicle density is d, then f signature operations and $f \times d$ signature verifications have to be performed per second. Moreover, as some applications need time-critical communication to some extent, the sum of both the time for creation and verification plays a role.

6.3.3 Secure Restricted Flooding/Geocast

Flooding is an approach that is used for a number of applications in VANETs to distribute information very quickly among the immediate surroundings of a vehicle. The basic principle involves multi-hop broadcast forwarding, which means that every node rebroadcasts the message once. As this cannot be done network-wide, the rebroadcast is usually restricted by either a time-to-live (TTL) counter value or a geographic destination area (GDA).

The purpose of this security component is to ensure integrity, authenticity and reliability of this mechanism. As a primary goal, the component is intended to prevent malicious vehicles being able to disturb the mechanism by means of rerouting, tampering and dropping. As a secondary goal, the module should be able to cope with attacks that intend to exploit the flooding mechanism to disturb the whole network operativeness. This is particularly important since flooding is a relatively costly mechanism that consumes a lot of bandwidth especially when node density is high.

Different actions need to be taken depending on whether a packet is incoming or outgoing. And in this case if it is created by the current node or forwarded only. Moreover, the applied security mechanisms partly depend on the mechanism used, i.e. whether the flooding restriction is TTL-based or GDA-based. An outgoing message may either originate from the current node or is to be forwarded by the current node. The required security processing differs notably.

For all messages created by one of the applications of node X, a signature has to be computed and a timestamp t_c has to be added if not already included. If the forwarding is TTL-restricted, then also a hash chain mechanism has to be applied, because malicious forwarders could decrease the TTL and thus increase the multi-hop propagation area. Such an increase leads, of course, to waste network bandwidth. If the restriction is given by a fixed geographic destination region, the hash chain is not necessary.

Hence, the first step is to include a timestamp t_c or to ensure that an accurate timestamp is already included. This is done together with the signature. The second step is to compute the hash chain in case of TTL-restricted forwarding. Therefore, the component has to generate a random base value v, apply a hash function TTLMAX times on it and append the result h_v as well as v to the message. As third step, the signature has to be created and the certificate for the used key (long term ID or pseudonym) has to be attached. For this step, it is important to distinguish between mutable and immutable fields (F_{m} and F_{im}). Fields like the TTL value or the hash chain base value v change during the forwarding, whereas other, immutable fields such as the payload, the source address or the end of the hash chain h_v does not change.

The signature should only be computed for these immutable fields, and not include mutable ones. For GDA-restricted forwarding, the message looks like this:

$$\begin{aligned} F_{\mathrm{im}} &= \mathrm{PAYLOAD}|X|\mathrm{GDA}|t_c \\ \mathrm{PACKETGDA} &= F_{\mathrm{im}}|sigSK_x(F_{\mathrm{im}})|certPK_x \end{aligned}$$

For TTL-restricted forwarding, the message includes the following:

$$\begin{aligned} F_{\mathrm{im}} &= \mathrm{PAYLOAD}|X|h_v|t_c \\ F_{\mathrm{m}} &= v|\mathrm{TTL} \\ \mathrm{PACKETTTL} &= F_{\mathrm{im}}|F_{\mathrm{m}}|sigSK_x(F_{\mathrm{im}})|certPK_x \end{aligned}$$

Packets forwarded by the local node need to be processed after the routing procedure. In particular, the hash chain base value v has to be replaced by

$h(v)$, i.e. the hash chain has to be shortened by one element, because the routing has decreased the TTL value.

Other fields, especially the signature and the immutable fields are not modified by forwarding nodes, but play a role to check incoming messages.

The primary purpose for the inspection of all incoming packets is checking security policies. One of these policies is the verification of the attached signature as well as the certificate. If the signature or the certificate cannot be verified, the message should be dropped. Moreover, more checks are necessary to ensure security. In summary, an incoming message should pass all the following checks before continuing processing (e.g. routing).

After receiving these messages, the receiver will execute certificate check, signature check, timestamp check, GDA size check, and hash chain check. If any of these checks fails, the message must not be forwarded. Regarding local reception, it is either discarded immediately or marked as invalid by the component. The choice depends on whether applications also want to process invalid packets and should be configurable.

Though these basic measures already can help against attackers, there are still some problems to be addressed:

(1) As soon as the notion of node location plays a role, there is always an attack opportunity against the positioning system that provides nodes with the current position. Thus, secure positioning could help for all position-related packet types in the network. If not all vehicles in a certain area are tricked in parallel (e.g. by a fake GPS satellite), also a heuristic approach to position verification can be helpful.

(2) Simple flooding and geocast mechanisms typically use broadcasts to send packets to all neighbors at once. Therefore, packets are not acknowledged by the receivers, which allow an attacker to selectively destroy packets on the data link layer. For a receiving node, the attack would just look like a collision which happens regularly in wireless ad hoc networks. Because both flooding and geocast generate a lot of redundancy if every intermediate node rebroadcasts a packet, this is not a problem in a large area where multiple paths exist and where an attacker only has a local impact. But, on highways, the radius of the transmission range is often enough to block all packets of one message from further forwarding. As there is no retransmission, these packets will get lost and the flooding/geocast ends there.

6.3.4 Secure Geographic Routing

With geographic routing, we denote multi-hop single-path forwarding method according to the principle of greedy geographic routing. The message's destination is a geographic coordinate rather than a node address. The basic concept of geographic forwarding is to pass messages always to a neighbor node, which is geographically closer to the destination than the current node.

To be able to select such a next hop for a packet, every node needs to know its one-hop neighbors and their current positions. The greedy geographic routing requires a periodic beaconing service to get the described neighbor information. More advanced mechanisms can work without beaconing. However, these mechanisms also have drawbacks, and as we need beaconing in VANETs anyway, we refer to the original form here.

The reason why this type of routing was favored over topological routing protocols for ad hoc networks like AODV or DSR is that has significant advantages in ad hoc networks with very high dynamics like it is the case in inter-vehicle networks.

To secure geographic routing, there are several aspects to be considered. Like in the previously described patterns, packets must be integrity protected and it is helpful to guarantee that packets can only be generated by legitimate participants of the network, such as registered vehicles or RSUs. This can be achieved by signing packets.

The more difficult aspects concern one of the building blocks of geographic routing, the beaconing. Apart from the general security considerations of beaconing, there are more problems to be solved with geographic routing.

6.4 Privacy Enhancing and Secure Positioning

6.4.1 Overview

Privacy preservation is an important design requirement for VANETs, where the source privacy of safety messages is envisioned to emerge as a key security issue because some privacy-sensitive information, such as the driver's name, license plate, vehicle model, position, and driving route, could be intentionally deprivatized so that the personal privacy of the driver is jeopardized. Thus, the safety message's authentication with source privacy preservation is critical for a VANET that is considered for practical implementation and commercialization. In particular, the privacy preservation in VANETs should be conditional, where senders are anonymous to receivers while senders should be traceable to the CA. The CA with the traceability can reveal the source identity of a message once a dispute occurs to the safety message.

In VANETs, position is one of the most important data for vehicles. Each vehicle needs to know not only its own position but also those of other vehicles in its neighborhood. The Global Positioning System (GPS) is the most widespread outdoor positioning system for mobile devices today. The system is based on a set of satellites that provide a three-dimensional positioning with an accuracy of around 3 m. However, GPS signals are weak, can be spoofed, and are prone to be jammed[17]. Moreover, vehicles can intentionally lie about their positions. Hence the need for a secure positioning system that will also support the accountability and authorization properties, frequently related to a vehicle's position.

6.4.2 Privacy Protection Enhancing Scheme

Some approaches have been proposed that claim to effectively provide privacy protection in Vehicle Communications (VCs). However, privacy requirements are often only implicitly stated. The explicit set of privacy requirements identified in section 1 allows us to assess the actual level of privacy protection achieved by an approach. VCs privacy approaches can be coarsely divided into five general categories; they are basic pseudonym approaches, extended pseudonym approaches, symmetric key approaches, group signature approaches, and IBC approaches. In reference [18], representative approaches from these categories are selected and how they fulfill the requirements are discussed.

6.4.2.1 Basic pseudonym approaches

In the context of VCs, pseudonyms commonly refer to pseudonymous public key certificates. These certificates are generated in a predefined way. They do not contain any identifiable information and cannot be used to link to a particular user or to another pseudonymous certificate. Vehicles are equipped with pseudonyms and their corresponding secret keys. When sending a message, a vehicle signs it with its secret key and attaches the signature and the pseudonym certificate to the message so that receivers can verify the signature. Vehicles also have to change pseudonyms often to make it hard for an attacker to link different messages from the same sender.

In reference [19], the SeVeCom project is proposed, which defines baseline security architecture for VC systems. Based on a set of design principles, SeVeCom defines an architecture that comprises different modules, each addressing certain security and privacy aspects. In privacy aspects, the SeVeCom approach employs a hierarchical CA structure, in which CAs manage and issue long-term identities to vehicles. Pseudonyms are issued by pseudonym providers and are only valid for a short period of time. When issuing pseudonyms, a pseudonym provider authenticates a vehicle by its long-term identity and keeps the pseudonyms-to-identity mapping in case of liability investigation. The secret keys of the pseudonyms are stored and managed by a Hardware Security Module (HSM), which is tamper-resistant to restrict the parallel usage of pseudonyms. Provided with a pseudonym, pseudonym resolution authorities can resolve an identity by accessing the pseudonyms-to-identity mappings at a pseudonym provider. Owing to the short lifetime of pseudonyms, the need for credential revocation is minimized. It is basic that only a vehicle's long-term identity is revoked to prevent it from acquiring new pseudonyms from a pseudonym provider. Consequently, CAs only need to distribute CRLs to pseudonym providers, which are part of the infrastructure network.

6.4.2.2 Extended pseudonym approaches

Approaches in this category aim to either improve or enhance specific aspects of the basic pseudonym approaches.

The PKI+ approach[20] is based on bilinear mappings on elliptic curves. It retains the concept of the well-known PKI approach, but provides the additional benefit. In the approach users are autonomous in deriving public keys, certificates and pseudonyms which minimizes the communication to the certificate authority. A user obtains a master key and certificate from a CA after it proves its identity and knowledge of a user secret x to the CA. The user can then self-generate pseudonyms by computing a public key from the master certificate, the secret x, and a random value. A certificate is computed as a signature of knowledge proof s over the public key and the master public key. The certificate also includes the version number *Ver* of the CA public key for revocation purposes. The user signs a message m by computing the signature of knowledge proof m_s on m. A receiver of m can verify the message with the public key in the pseudonym. When revoking a user, the CA publishes a new version information Ver', which has to be used by all users to update their keys. Ver' is chosen so that it is incompatible with the master key and master certificate of the revoked user. The advantage of the PKI+ approach is that vehicles do not need to contact a CA or pseudonym provider to obtain new pseudonyms. The disadvantages of the approach are that Sybil attacks based on unlikable pseudonyms are hard to detect and that the CA has no means to control the amount of self-generated pseudonyms.

The blind signature approach[21] applies blind signatures and secret sharing in the pseudonym issuance protocol to enforce distributed pseudonym resolution. In the approach, a user blinds the public key to be signed and presents shares of it to a number of CAs in the pseudonym issuance process. Each CA is possessed of a partial secret of the secret key shared by all CAs in a secret sharing scheme. Each CA signs the presented blinded key part with its partial secret key, returns it to the user, and stores a corresponding partial resolution tag in its database. The user can unblind and combine the received results, yielding a certificate which can be verified with a public key commonly to all CAs. The certificate is only valid if k of n CAs participated in the issuance process. Otherwise the threshold of the secret sharing scheme is not reached, thus resulting in an incomplete signature. To resolve a pseudonym, more than t CAs have to cooperate in a second secret sharing scheme to compute a joint resolution tag for the presented pseudonym and compare it to all tags in the database. The advantage of the scheme is that it effectively prevents misuse of resolution authority. The disadvantage of the scheme is that it incurs considerable overhead by requiring a number of authorities to take part in the certification of a single pseudonym. In addition, pseudonym resolution requires comparisons with all tags stored in the revocation database, and consequently, does not scale well with the number of users.

6.4.2.3 Symmetric key approaches

Symmetric cryptography schemes are more efficient for time-critical applications for the reason that symmetric cryptography schemes require less computational effort than asymmetric operations. However, symmetric encryption has to somehow emulate asymmetric properties in order to achieve authentication.

The TESLA approach[22] is based on the TESLA lightweight broadcast authentication mechanism[23]. TESLA uses time as the creator of asymmetric knowledge to create asymmetric properties similar to public key cryptography, assuming that network nodes are loosely synchronized. Time synchronization requirements for VANET nodes are to be feasibly given by current technology. In the scheme, a user computes a key chain and releases keys subsequently in fixed time intervals. Each message is authenticated with a key that has not yet been released according to the key schedule, and receivers have to buffer messages until the corresponding key is released and the message can be verified. The authenticity of a message can be verified with any key higher up in the chain. The advantage is that TESLA keys are much shorter in length than public keys and are thus more efficient. To enhance trust, each vehicle also has a set of pseudonyms signed by a CA. Pseudonyms are only used to sign anchors of the key chains. When two vehicles enter each other's reception range, they first exchange certificates to obtain each other's TESLA anchors. Subsequently, they only use symmetric TESLA keys to authenticate messages. Keys belonging to the same key chain as the presented anchor can be traced back to it and thus verified. The proposed scheme significantly reduces the security overhead comparing to the current DSRC draft standard on security (IEEE P1609.2). It provides efficient authentication while reducing certificate exchanges to a minimum. However, for time-critical safety applications the delay in authentication may create problems. Otherwise, in the scheme the keys expire too quickly and actual receivers might not receive disclosed keys. Therefore, TESLA keys are not suitable for multi-hop forwarding.

6.4.2.4 Group signature approaches

A Group signature scheme is a method for allowing a member of a group to anonymously sign a message on behalf of the group. It provides conditional anonymity to members of a group. Each group member can create signatures which can be verified with a common group public key. Essential to a group signature scheme is a group manager, who is in charge of adding group members and has the ability to reveal the original signer in the event of disputes. Only the group manager is able to determine the identity of a signer.

The hybrid approach[24] uses group signatures to reduce the overhead of key and pseudonym management. Vehicles are members of a group and equipped with a secret group signing key and the group public key. Each vehicle generates random public/secret key pairs to be used for pseudonymous communications. The public keys are signed with the group secret key, yielding a pseudonym certificate that can be verified with the group public key. When communicating, vehicles sign the outgoing messages with the secret key of the pseudonym and attach the pseudonym to the message. Upon receipt of such a message, a receiver can verify that the pseudonym was created by a legitimate group member with the group public key. When necessary the group manager is able to open group signatures and retrieve the signer's identity. The scheme enables vehicle on-board units to generate their own pseudonyms, without affecting the system security. One advantage of the scheme is that it obviates the need to acquire new pseudonyms periodically. However, in the scheme revocation of group membership is a scalability issue nevertheless.

The GSIS approach[25] is based on short group signatures and identity-based signature techniques. In the approach, a CA acts as the group manager and has the ability to reveal the original signer. The CA computes a group public key and group secret keys for each vehicle in the group from their unique identifiers. With the identifier and a part of the secret key, a CA can determine the identity of a group member. Therefore accountability can be achieved while at the same time impersonation attacks are prevented. Similar to the hybrid approach, a vehicle signs messages with its own secret key and receivers can verify them with the group public key. Revocation is achieved by distributing revocation lists. One difference to other schemes is that revocation lists are only allowed to grow to a threshold t to avoid increasing verification times. When t vehicles have been revoked, the group key and individual secret keys are updated. The disadvantage of this scheme is that the CRL may grow quickly, which may not only have a large CRL size, but also take a long time to look through the whole CRL to see if a certificate is still valid or not.

6.4.2.5 IBC approaches

Identity-based cryptography (IBC) is a type of public-key cryptography in which a publicly known string representing an individual or organization is used as a public key. The public string is the individual's (or organization's) identity that could include an email address, domain name, or a physical IP address. Presented with a signature, a verifier can check its validity merely by knowing the sender's identity.

The efficient conditional privacy preservation (ECPP) approach[26] utilizes both IBC and group signatures. In the scheme, a trusted authority TA sets up an IBC scheme and publishes its system parameters. Each vehicle has a unique identity, which is used to authenticate with the TA to obtain a pseudonym. When a vehicle submits its identity, TA generates a pseudo identifier by encrypting the vehicle identifier with its public key and extracting a corresponding private key from it. The vehicle can use the resulting key pair as a pseudonym in anonymously authentication processes with RSUs under control of TA. When a vehicle enters the vicinity of a RSU, it requests a short-time anonymous key certificate to take part in a local group signature scheme. For this reason, the group identifier is also used as the group public key. The RSU checks that the presented pseudonym is not listed on a CRL, and issues a group membership certificate, which is valid only for a short period of time. The RSU also retains a mapping between group membership certificate and pseudonym. Where after, the vehicle can perform group signatures on messages by proving possession of a membership certificate, and therefore communicate anonymously with other vehicles. By opening the group signature of a message and retrieving the identifier of the RSU that issued the group membership certificate, the TA is able to realize identity resolution. The RSU can then be contacted and returns the pseudonym corresponding to the presented membership certificate. In the last step, the TA decrypts the pseudonym with the symmetric key and yields the real vehicle identifier. The advantage of the scheme is that it can provide fast anonymous authentication and privacy tracking while minimizing the required storage for short-time anonymous keys.

6.4.3 Secure Positioning Scheme

Secure positioning will play a significant role in many vehicular applications, making it critical to determine that a message did indeed originate at a given location. For example, secure positioning would prevent an attacker sitting on the side of the road from claiming to be a vehicle traveling on the highway. It would also prevent an adversary from using another communication medium to replay a message heard in one location as though it had originated in a different location.

GPS-based positioning has a lot of disadvantages. It cannot be used for indoor positioning or for positioning in dense urban regions: in those cases, because of the interferences and obstacles, satellite signals cannot reach the GPS devices. Furthermore, civilian GPS was never designed for secure positioning. Civilian GPS devices can be spoofed by GPS satellite simulators, which produce fake satellite radio signals that are stronger than the real signals coming from satellites. Until now, there is little work done on secure positioning without GPS. One possible approach would be to extend the protocols that have been proposed for secure localization in sensor networks to this new setting. Unfortunately, these protocols such as reference [27 – 29] focus on allowing a sensor to securely determine its own position (rather than the positions of its neighbors) or rely on the presence of multiple base stations. In reference [30], Parno et al. propose to leverage the properties of the vehicular environment to provide a new method of secure relative localization. In their scheme, a vehicle's relative location is defined by its entanglement with other vehicles. Each vehicle will regularly broad cast its identity (a public key) along with its signature of a current timestamp. When a vehicle receives such a broadcast, it signs the other vehicle's ID and rebroadcasts it. In other words, when vehicle A receives public key K_B from vehicle B, it adds a signature $\{K_B\}_{SK_A}$ with its private key SK_A to its regular broadcast. When vehicles pass each other traveling in opposite directions, this will allow both streams of traffic to perform relative localization (see Fig. 6.9). If vehicle B hears vehicle C rebroadcast A's identity before it rebroadcasts B's identity, then B can conclude that A is ahead of him/her. Vehicle B can aggregate multiple indicators (i.e., from vehicles D and E) to provide further assurance of A's position. Furthermore, vehicle B can evaluate the entanglement data for those vehicles as well to determine how much weight to give their reports.

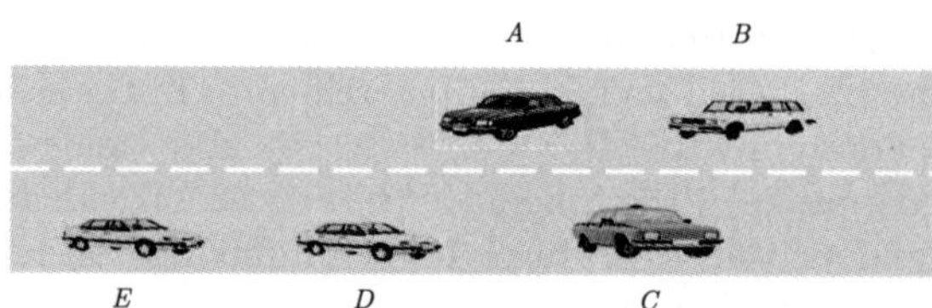

Fig. 6.9 Secure relative localization (vehicle B can use broadcasts from vehicles C, D and E to determine A's location).

This scheme helps to perform relative localization. But this approach incurs overhead and does not provide absolute positions. The final solution will probably be a hybrid system that will use a combination of GPS, radars, wheel rotation sensors, digital maps, and roadside beacons, depending on the availability and reliability of each of these techniques.

6.5 Conclusion

The area of vehicular ad hoc networks has been developed significantly during the past decade. Several new applications are enabled by this new kind of communication network. However, as those applications have impact in road traffic safety, strong security requirements must be achieved. New mechanisms have to be developed to deal with the inherent features of these networks (extreme node's speed, decentralized infrastructure, etc.). In this chapter, we present an overview of the current security issues over VANETs. We have identified the security and privacy requirements and security threats in VANETs. We have also described security architecture for VANETs based on the PKI and security hardware, and introduced a secure VANET communication scheme based on TPMs. Furthermore, we have presented the scheme of key management and authentication under the security architecture based on the PKI and security hardware. Security routing solutions for V2V, V2I, and group communications have also been analyzed representatively. Finally, we have discussed privacy enhancing and secure positioning schemes.

Secure communications for vehicular ad hoc networks have become an important research issue these years. Several future research lines can be pointed out in VANET security area. Although several mechanisms have been proposed, some issues still have to be addressed (e.g. privacy problems due to radio frequency fingerprinting). Simulation results are often offered to evaluate current proposals. However, a common scenario to evaluate alternatives does not exist. Finally, hardware implementation of efficient cryptographic primitives is required in vehicles. In this way, achieving computation availability would be eased[31].

References

[1] IEEE Draft Std P802.11p/D2.0 (2006) Wireless Access in Vehicular Environments (WAVE).

[2] IEEE Std 1609.2-2006 (2006) IEEE Trial-Use Standard for Wireless Access in Vehicular Environments-Security Services for Applications and Management Messages. IEEE, New York.

[3] Luo J, Hubaux J P (2004) A Survey of Inter-Vehicle Communication, EPFL Technical Report IC/2004/24. http://infoscience.epfl.ch/record/52616/files/IC_TECH_REPORT_200424.pdf. Accessed 10 October, 2011.

[4] Sichitiu M L, Kihl M (2008) Inter-vehicle communication systems: a survey. IEEE Communication Surveys and Tutorials, 10(2): 88 – 105.

[5] Hartenstein H, Laberteaux K P (2008) A tutorial survey on vehicular ad hoc networks. IEEE Communications Magazine, 46(6): 164 – 171.

[6] Camp T, Boleng J, Davies V (2002) A Survey of Mobility Models for Ad Hoc Network Research. Wireless Commun. & Mobile Comp., special issue on Mobile Ad Hoc Networking: Research, Trends and Applications, 2(5): 483 – 502.

[7] Hubaux J P, Capkun S. Jun L (2004) The security and privacy of smart vehicles. IEEE Security and Privacy magazine, 2 (3), 49 – 55.

[8] Raya M, Hubaux J P (2007) Securing Vehicular Ad Hoc Networks. Journal of Computer Security, Special Issue on Security, Ad Hoc and Sensor Networks, 15(1): 39 – 68.

[9] Raya M, Papadimitrators P, Hubaux J P (2006) Securing vehicular communications. Wireless Communications, 13(5): 8 – 15.

[10] Guette G, Bryce C (2008) Using TPMs to secure ad hoc networks. In: Proceedings of the 2nd IFIP WG 11.2 international conference on information security theory and practices: smart.devices, convergence and next generation networks, pp. 106 – 116.

[11] Trusted Computing Group (2007) TPM main specification. Main Specification Version 1.2 rev. 103, Trusted Computing Group.

[12] Raya M, Hubaux J P (2007) Securing vehicular ad hoc networks. Journal of Computer Security, 15: 39 – 68.

[13] Wohlmacher P (2000) Digital certificates: a survey of revocation methods. In: Proceedings of ACM Wksp. Multimedia, pp. 111 – 114.

[14] Lin X D, Lu R X, Zhang C X, Zhu H J, Ho P H (2008) Security in vehicular ad hoc networks. IEEE Communications Magazine, 46(4): 88 – 95.

[15] Zhang C, Lin X D, Lu R X, Ho P H (2008) An efficient RSU-aided message authentication scheme in vehicular communication networks. In: Proceedings of the IEEE Conference on Communications, pp. 1451 – 1457.

[16] Antonio Kung (2008) Security Architecture and Mechanisms for V2V / V2I, D2.1 v3.0. http://www.sevecom.org/Pages/ProjectDocuments.html. Accessed 10 October, 2011.

[17] Capkun S, Hubaux J P (2006) Secure positioning in wireless networks. IEEE Journal on Selected Areas in Communications, 24(2): 221 – 232.

[18] Schaub F, Ma X D, Kargl F (2009) Privacy requirements in vehicular communication systems. In: Proceedings of the International Conference on Computational and Engineering, pp. 139 – 145.

[19] Papadimitratos P, Buttyan L, Holczer T (2008) Schoch E, Freudiger J, Raya M, Ma Z, Kargl F, Kung A, Hubaux J P, Secure vehicular communications: Design and architecture. IEEE Communications Magazine, 46(11): 100 – 109.

[20] Armknecht F, Festag A, Westhoff D, Zeng K (2007) Cross-layer privacy enhancement and non-repudiation in vehicular communication. In: Proceeding of the 4th Workshop on Mobile Ad Hoc Networks, pp. 1 – 12.

[21] Fischer L, Aiijaz A, Eckert C, Vogt D (2006) Secure revocable anonymous authenticated inter-vehicle communication. In: Proceeding of the 4th Workshop on Embedded Security in Cars (ESCAR).

[22] Hu Y C, Laberteaux K P (2006) Strong VANET security on a budget. In: Proceeding of the 4th Workshop on Embedded Security in Cars (ESCAR).

[23] Perrig A, Canetti R, Tygar J D, Song D (2002) The TESLA broadcast authentication protocol. RSA Cryptobytes, 5(2): 2 – 13.

[24] Calandriello G, Papadimitratos P, Hubaux J P, Lioy A (2007) Efficient and robust pseudonymous authentication in VANET. In: Proceeding of the 4th ACM Intl workshop on Vehicular ad hoc networks, pp. 19 – 28.

[25] Lin X, Sun X, Ho P H, Shen X (2007) Gsis: A secure and privacy-preserving protocol for vehicular communications. IEEE Trans. Vehicular Technology, 56(6): 3442 – 3456.

[26] Lu R, Lin X, Zhu H, Ho P H, Shen X (2008) Ecpp: Efficient conditional privacy preservation protocol for secure vehicular communications. In: Proceeding of the 27th Conference on Computer Communications, pp. 1229 – 1237.

[27] Sastry N, Shankar U, Wagner D (2003) Secure verification of location claims. In: Proceedings of WiSe, pp. 1 – 10.

[28] Ji X, Zha H Y (2004) Sensor positioning in wireless ad hoc sensor networks using multidimensional scaling. In: Proceeding of the 23rd conference of the IEEE Computer and Communications Societies, pp. 2652 – 2661.

[29] Bras L, Oliveira M, Carvalho N B, Pinho P (2010) Low power location protocol based on ZigBee wireless sensor networks. In: Proceeding of the international conference on Indoor Positioning and Indoor Navigation, pp. 15 – 17.

[30] Parno B, Perrig A (2005) Challenges in securing vehicular networks. In: Proceedings of the Workshop on Hot Topics in Networks.

[31] de Fuentes J M, Gonzalez-Tablas A I, Ribagorda A (2010) Overview of security issues in vehicular ad hoc networks. Handbook of Research on Mobility and Computing. IGI Global, Hershey.

Chapter 7
Security in Wireless Sensor Networks

Weiping Wang[1], Shigeng Zhang, Guihua Duan, and Hong Song

Abstract

Wireless sensor networks (WSNs) are exploiting their numerous applications in both military and civil fields. For most WSNs applications, it is important to guarantee high security of the deployed network in order to defend against attacks from adversaries. In this chapter, we survey the recent progress in the security issues for wireless sensor networks, mainly focusing on the key distribution and management schemes and some high layer protocols such as secure routing, location privacy protection and secure data aggregation. Representative works on each topic are described in detail and both of their strongpoint and drawbacks are discussed, based on which we give some direction for future research.

7.1 Introduction

Wireless sensor networks (WSNs) are a type of Ad Hoc networks that consist of a large number of resource-constrained sensor nodes. Requiring no fixed infrastructure, WSNs can be quickly deployed, organized, and maintained in an ad hoc manner. The flexibility in deployment and maintenance advances WSNs' application in many fields, including military, environmental monitoring, public safety monitoring, emergency handling, medical, and oceanic monitoring. For example, WSNs can be used to detect and track the intrusion of enemies or their tanks on a battlefield, to detect forest-fires and floods, to monitor environmental pollutions, or to measure traffic flows in a traffic network.

Security is one of the most important issues in WSNs. As WSNs are usually deployed in hostile or remote environments and work in an unattended

1 School of Information Science and Engineering, Central South University, Changsha, 410083, China.

manner, prevention of network attack by adversaries and protection of privacy of sensitive collected data is pivotal for many WSNs applications. However, it is challenging to provide security in WSNs due to the following reasons.

1. *Large scale deployment*

The number of sensor nodes in a WSN can be very large, sometimes maybe several orders of magnitude larger than that of a traditional Ad Hoc network. Furthermore, in order to provide redundancy, sensor nodes are usually densely deployed in the target area. These two factors necessitate good scalability of security protocols designed for WSNs.

2. *Extremely limited resources of sensor nodes*

Sensor nodes are usually extremely resource-constrained, e.g., in communication bandwidth and power supply. Thus we need to keep in mind energy efficiency and low cost when designing protocols for WSNs.

3. *Dynamic network topology*

The topology of a WSN may change frequently due to many factors after deployment, e.g., node failures, new node deployment, old node revocation, or node movements. Coping with the changes in topology will greatly impact the performance of security protocols.

4. *Lack of global identifications*

In contrast to traditional IP-based networks, nodes in WSNs have no global identifications. This prohibits application of many existing security protocols that rely on unique identification of nodes within the network. New security mechanisms that do not require nodes' global identifications need to be designed.

Table 7.1 lists typical attacks that can be launched in WSNs for different network layers. Besides some traditional security threats such as information disclosure, tampering, replay attack and denial of service, there are some new attacks in WSNs including sink node attack, Sybil attack, sinkhole attack, node replication attack, random walk attack and wormhole attack. Because of the lack in network infrastructures and the hostile deployment environments, WSNs are very susceptible to these new attacks.

Table 7.1 Main security threats in WSNs

Layer	Security Threats
Physical	Jamming Attack, Physical Tampering Attack
Data Link	Collision Attack, Exhaustion Attack, Unfair Competition Attack
Network	False Routing Information Attack, Selective Forwarding Attack, Sinkhole Attack, Sybil Attack, Wormhole Attack, HELLO Flood Attack, Acknowledgment Spoof Attack, Passive Wiretapping Attack
Transport	Flooding Attack, Desynchronization Attack

Currently, research on security in WSNs mainly focuses on key management, secure routing, secure data aggregation and position privacy protection of key nodes in the network. Key management is discussed in Section 7.2. It is the basis of other security protocols, which establishes pair-wise session keys between nodes to provide secure communication links. Secure routing protocols provide safe end-to-end data delivery in WSNs. Typical secure routing protocols are discussed in Section 7.3. In Section 7.4, we address the problem of how to protect location privacies of key nodes, including both the sink node and some source nodes, and survey recent progress. Data aggregation provides energy-efficient approaches to collecting data in WSNs. However, aggregation requires data to be interpreted in intermediate nodes, which may result in data exposure. This conflicts with the goal of security that requires data encryption between nodes. We give a comprehensive overview of existing secure data aggregation protocols in Section 7.5.

7.2 Key management in WSNs

Key management is the basis to build secure WSNs. In order to satisfy security requirements such as confidentiality, integrity, node authentication and network availability, data transmission in WSNs needs to be encrypted or authenticated with keys. This requires establishment of secure communication links between nodes, which consequently requires effective key distribution and management.

7.2.1 Classification of key management schemes

There are two kinds of keys used to build secure communication links in WSNs: the initial keys and the session keys. The initial keys are generated and pre-stored on sensor nodes before they are deployed and are usually used to generate session keys. An initial key can be a shared key, a (or a set of) key parameter(s), or a key ring composed of a key chain. The session keys are usually generated after the deployment of nodes using initial keys; they are the actual keys that are used in establishing secure communication links between nodes.

The cryptography systems used to encrypt messages in WSNs can be either symmetric or asymmetric[1]. Compared with asymmetric cryptography systems, symmetric cryptography systems use keys with shorter length and usually incur less computational overhead; but the management and distribution of keys with symmetric cryptography are relatively more complex. On the other hand, key management and distribution in asymmetric cryptography systems are simple, but asymmetric cryptography systems usually incur high computational overhead and require hardware with strong ability. As a result of these limitations, symmetric cryptography systems are more suitable for current WSNs. Thus most existing key management protocols designed for WSNs use symmetric keys.

A key management protocol needs to provide functions including initial keys pre-distribution, session keys generation, and key updating. Due to the extremely limited resources of sensor nodes, we should keep in mind reduction of computational and communication overhead when designing key management protocols for WSNs. On the other hand, due to the open and hostile environments where WSNs are usually deployed, the designed protocols should be robust, and resistant to unexpected node failure or intentional attacks launched by adversaries. It should be guaranteed that the exposure of keys of a compromised node would not affect the security of communications among other nodes.

The topology of a WSN cannot be known before its deployment. Thus the session keys can only be generated after deployment based on pre-distributed initial keys via negotiations among nodes. According to the schemes used to distribute initial keys and the approaches used to negotiate between nodes, key management mechanisms in WSNs can be roughly classified into three categories.

1. *Centralized schemes*

Centralized schemes usually rely on a trusted key distributed center (KDC) to distribute and manage keys, e.g., the SPINS protocol[2]. In centralized schemes, shared keys between nodes and the sink node are pre-loaded into sensor nodes before deployment, and the session keys' generation and update are both conducted by the sink node. The advantage of centralized key management schemes is that they require only small memory space and low computational capability for sensor nodes to get keys; the disadvantage is that they incur high communication overhead in generating and updating keys via negotiation. Furthermore, centralized schemes overly rely on the sink node to manage keys, thus are very vulnerable to the single point of failure: If the sink node is compromised, the entire network will be under threat.

2. *Distributed schemes*

In contrast to centralized schemes, in distributed key management schemes there are no KDCs and the generation and update of keys are performed in completely distributed manners. Before deployment, a setup server preloads initial keys or some parameters for generating keys into sensor nodes. Then, after deployment, sensor nodes generate and update their session keys by themselves with the initial keys or parameters. This type of scheme is also referred to as a key pre-distribution scheme (KPS). Typical KPSs include E-G[3] and Q-composite[4]. The advantage of distributed key management schemes is that the generation and update of session keys are performed in purely distributed manners, thus there is no single point of failure problem as in centralized schemes. This makes the network as a whole, hence more robust. The disadvantages are that it requires sensor nodes to store a large amount of preloaded keys which incurs high storage and communication overhead.

3. *Hierarchical schemes*

Hierarchical key management schemes make tradeoffs between centralized schemes and distributed schemes and leverage advantages of both of these schemes. Examples of hierarchical key management schemes include LEAP[5] and LOCK[6]. In hierarchical schemes, nodes in the network are divided into different clusters; in each cluster there is a cluster head which manages the generation and update of keys for cluster members. Research on hierarchical key management schemes is a hot spot in recent years.

7.2.2 Two well-known key management schemes

In this section we introduce two well-known key management schemes for WSNs: the Blom scheme and the Blundo scheme.

1. *The Blom scheme*

The Blom scheme proposed in reference [7] exploits the characteristics of symmetric matrices to generate pair-wise session keys for neighboring nodes in the network. There are two matrices used in the Blom scheme: a public $(\lambda+1)\times N$ matrix G and a secret random symmetric $(\lambda+1)\times(\lambda+1)$ matrix D. Let $A=(DG)'$ and $K=AG$, then K is also symmetric. Before deployment, the k-th node is assigned with the k-th row of $A(A_k)$ and the k-th column of $G(G_k)$ and is stored in memory. After the deployment, two neighboring nodes exchange their column vectors and compute their secret session key. For example, node i and node j exchange their column vectors G_i and G_j and compute their symmetric secret session key as $K_{i,j}=A_i\times G_j=A_j\times G_i=K_{j,i}$.

The Blom scheme is λ-secure, which means that the network is perfectly secure as long as there are no more than λ compromised nodes. Thus this scheme is not vulnerable to the single point of failure problem if $\lambda>1$. Furthermore, when λ equals to the number of nodes n, the network will be perfectly secure. However, in a λ-secure Blom scheme each node needs $O(\lambda+1)$ memory space to store the keys and corresponding vectors, which incurs high storage and computational overhead.

2. *The Blundo scheme*

Blundo et al.[8] propose a scheme to generate pair-wise session keys for nodes based on symmetric binary polynomials. In this scheme, a setup server first generates a symmetric λ-order bi-variable symmetric polynomial $f(x,y)=\sum\limits_{i,j=0}^{\lambda}a_{ij}x_iy_j$ that satisfies $f(x,y)=f(y,x)$, then assigns the node with ID i a polynomial share $f(i,y)$. After the deployment, two nodes i and j exchange their IDs and calculate their shared key with $K_{i,j} = f(i,j) = f(j,i) = K_{j,i}$.

The Blundo scheme is considered to be a special polynomial implementation of the Blom scheme. It can provide the same λ-security as the Blom scheme.

7.2.3 Typical centralized schemes

In centralized key management schemes, the sink node plays the role of a trusted KDC. Every node i in the network shares an initial key K_i with the KDC. The communication key between two nodes is generated and updated via negotiations between the node and the sink nodes. The messages in the negotiation process are encrypted or authenticated using corresponding K_i.

In this kind of scheme, each node only requires small memory space to store its keys, and the computational overhead in calculating pair-wise session keys is low. However, the over-reliance on the interactions with the KDC to generate and update session keys makes centralized schemes low scalable as well as vulnerable to single point of failure. If the KDC is compromised, the security of the entire network will be destroyed.

Because of their poor scalability, centralized schemes are not suitable to large scale WSNs. Up to now, only a small number of security protocols proposed for WSNs use centralized schemes, among which the most famous one is the *Security Protocols for Sensor Networks* (SPINS) proposed by Perrig et al.[2].

In SPINS, each node shares an initial key with the sink node called the *master key*. There are three keys for every node in SPINS, namely the encryption key K_{enc} used to encrypt messages exchanged between the node and the sink, the message authentication key K_{mac} for message authentication and K_{rand} used to generate pseudo random numbers. They are all derived from the master key.

SPINS includes two sub-protocols: a *Security Network Encryption Protocol* (SNEP) and a *Timed Efficient Stream Loss-tolerant Authentication* protocol (μTESLA). SNEP provides confidentiality, integrity and freshness of transmitted messages in point to point communications. In the communication process, the encryption key K_{enc} is used to encrypt exchanged messages between two nodes which provide message confidentiality, while the authentication key K_{mac} is used to authenticate messages which provide message integrity. In order to provide data freshness and resist replay attacks, SNEP uses different random numbers for different message transmissions. SNEP also provides semantic security. A piece of plaintext will be encrypted into different ciphertexts at different time. This is achieved by preceding a message with a random bit string before encrypting the message with an encryption function. The random bit string is necessary for the receiver to decrypt the received message. A shared counter between the sender and the receiver is used to generate the random bit string rather than directly transmit it, which avoids incurring additional communication overhead.

Denoted by $MAC(K, x)$, the process of generating MAC for message x with revision key K. Then the process of generating the session key between node A and node B can be described as

$$\begin{aligned}
&A \to B : N_A, A \\
&B \to S : N_A, N_B, A, B, MAC(K^B_{\text{enc}}, N_A|N_B|A|B) \\
&S \to A : \{K_{AB}\}_{K^A_{\text{enc}}}, MAC(K^A_{\text{MAC}}, N_A|B|\{K_{AB}\}_{K^A_{\text{enc}}}) \\
&S \to A : \{K_{AB}\}_{K^B_{\text{enc}}}, MAC(K^B_{\text{MAC}}, N_B|B|\{K_{AB}\}_{K^B_{\text{enc}}})
\end{aligned}$$

where N_A and N_B are random numbers generated by A and B to ensure data freshness, K_{AB} is the session key between A and B generated by the sink node. $\{M\}_K$ means encrypting message M with key K.

Sometimes the sink node needs to broadcast messages into the whole network, e.g., broadcasting which node is unavailable. In order to guarantee message confidentiality, the sink node would first encrypt the message with a chosen key and broadcast the ciphertext to other nodes. It then reveals the decryption key to all the nodes in the network. The μTESLA protocol is used to perform broadcast authentication and ensures the messages are indeed sent by the sink node.

The μTESLA protocol uses a symmetric mechanism to authenticate broadcast messages. Firstly, the sink node uses a one-way hash function H to generate a MAC key chain $\{K_0, K_1, \cdots, K_n\}$, in which $K_i = H(K_{i+1})$. The key chain has the property that it is easy to calculate $K_i, \cdots, K_0$ given K_{i+1} while the opposite is difficult. The time is divided into discrete time intervals; in each time interval there is a key in the key chain to be used. In the j-th time interval, the sink node authenticates messages with K_j but delays the announcement of K_j by a time of δ. Upon receiving a broadcasted message from the sink node in the j-th time interval, a node buffers the message in its memory and waits for the exposure of the corresponding key K_j. After receiving K_j, the node first authenticates the legitimacy of K_j using a previously stored key K_i by checking if $K_i = H^{j-i}(K_j)$ holds. If K_j is illegal, the node then uses K_j to authenticate the previously buffered packets.

The SPINS protocol only needs small memory to store keys, thus it incurs low storage overhead. Because every node independently shares its key with the base station, SPINS provides good resistance to node capture attacks. The μTESLA protocol uses a symmetrical mechanism for broadcast authentication, thus the energy consumption is low. However, in SPINS the key negotiation and data authentication of all nodes in the network are performed by the sink node, which will make the entire work under threat once the sink node is compromised. Furthermore, the traffic load at the sink node is very high and proportional to network size which limits its use in large scale sensor networks.

7.2.4 Typical distributed schemes

In distributed schemes, session keys are generated in a full distributed manner and do not rely on the sink node. Thus distributed key management schemes have good scalability and are more suitable for self-organized WSNs. Among others, the Key Pre-distribution Scheme (KPS)[3] is a typical distributed scheme. There are three phases in KPS:

(1) Key pre-distribution, in which some initial keys are pre-loaded into nodes before they are deployed.

(2) Shared keys discovering, in which neighboring nodes exchange their identify information and calculate their session keys using pre-loaded keys.

(3) Path key setup, in which two nodes establish session keys indirectly using some intermediate nodes with which they both share keys in case they failed to generate session key in the second phase.

There are a variety of methods to implement key pre-distribution schemes. The extremely simple method is to let all the nodes in the network share an identical master key; each pair of nodes uses this master key to generate their session key. With this method, every node only needs very small space to store its keys, thus incurs low storage overhead. The disadvantage is that if only one node is compromised, the security of the entire network is ruined. Another extreme method, however, is to let every node store its shared keys with all other nodes (the shared keys are diverse for different nodes) and to generate pair-wise session keys with different shared keys. This method provides the highest security in the means that a compromised node does not affect secure communications among other nodes, but it incurs high storage overhead on sensor nodes.

Current research on key pre-distribution schemes mainly focuses on how to reduce communication/storage overhead in order to save energy consumption, how to improve the scalability, and how to improve the ability to resist node capture attacks.

7.2.4.1 The E-G protocol

The E-G protocol[3] uses random key pre-distribution to establish shared keys between two nodes. It uses a key pool to pre-load keys to nodes before deployment. A key pool includes P different keys and each key is associated with a key ID. Before a node is deployed, $k(k \ll P)$ randomly selected keys from the key pool are pre-loaded into the memory of that node, which is called the key ring of that node. The values of k and P are carefully selected such that two adjacent nodes have shared keys with a probability p larger than a given threshold, where p is defined as

$$p = 1 - \frac{[(P-k)!]^2}{(P-2k)!P!}$$

After deployment, two neighboring nodes exchange their key IDs and find their shared keys by comparing their key rings in a shared-key discovery phase. If two nodes share more than one key, they randomly chose one as their session key. If there is no shared key between two nodes, they need to find an intermediate node between them with which they both share a key and establish their session key indirectly based on that key. This phase is called the path key establishment phase.

In the E-G protocol, the session keys between nodes are established using pre-loaded keys and need no negotiation or interaction with the sink node. Thus it is easy to implement and provide high scalability and flexibility. Each node needs to store k initial keys and exchange its key ring with neighbors, which incurs higher storage and communication overhead compared with the centralized schemes. Furthermore, the probabilistic mechanism used in E-G makes it fail to guarantee the existence of common keys between any two adjacent nodes; there may be some neighboring nodes that have no common keys. When two nodes have more than one common key, they will randomly choose one as their session key. This makes it possible that different node pairs use a same session key, which degrades the protocol's ability to resist node capture attacks.

E-G is the first protocol that uses random key pre-distribution to perform key generation and update in WSNs. Many follow-up schemes are proposed based on the idea of E-G which tries to enhance its security or reduce its storage/communication overhead by tuning the threshold on the number of common keys, using different methods to generate the key pool, or adopting different key pre-distribution methods. We will describe them as follows.

7.2.4.2 *Q*-composite: Enhancing security of E-G

The Q-composite scheme proposed by Chan et al.[4] tries to enhance the security level of the E-G scheme by using more common keys to generate session key. In the E-G scheme, it is only required that two neighboring nodes share at least one common key to establish their session key. In the Q-composite scheme, it is required that two neighboring nodes share at least $q > 1$ common keys to establish their session key. Furthermore, as opposed from the E-G scheme in which a randomly selected common key is directly used as the session key, in the Q-composite scheme the session key is generated using all common keys with a hash function, which reduces the probability that different node pairs in the network use same session keys and consequently increases the network's ability against node capture attacks.

The basic idea of Q-composite is as follows. In the initialization phase, for each node a set of m random keys are picked out of a key pool with total $|S|$ keys and are stored into that node's memory. In the key-setup phase, only the neighboring nodes sharing more than q keys can establish their session keys. Assume two nodes share t keys where $t \geqslant q$. Then the session key for the two nodes is established as $K = H(k_1||k_2||\dots||k_t)$, where H is a public hash function known to all nodes.

Generally speaking, the security level increases if larger threshold q is used in the Q-composite scheme. This is because when q increases, the number of keys used to generate session key between two nodes also increases, which consequently makes it more difficult for an adversary to successfully attack the session key. However, larger q may also make the network more vulnerable: If the adversary compromises a node, it also obtains at least q keys that node shares with its neighbors. When the number of compromised nodes is large, the security of Q-composite may be even weaker than that of the original E-G scheme. Furthermore, larger q requires larger memory space. Thus we can adjust the value of q to make a tradeoff between the security level and the storage overhead.

7.2.4.3 DDHV and RS: enhancing key connectivity of E-G

In E-G and Q-composite, each node is pre-loaded with some initial keys chosen from a key pool and two nodes establish their session key with a probability. Aiming to improve the key connectivity (in terms of number of session keys between two nodes) of the network, some protocols are proposed, e.g., the DDHV scheme proposed by Du et al.[9] and the RS scheme proposed by Liu et al.[10].

1. *DDHV*

Du et al. propose a multiple-space key pre-distribution scheme based on Blom's work, which we called DDHV in this paper. The Blom scheme uses symmetric matrix to generate session keys for every node pair in the network. Du et al. modify the structure of the key pool in E-G and use symmetric matrices to generate session keys for nodes in the network, which increases the possibility of establishing session keys for any two nodes. Furthermore, DDHV uses multiple key pools rather than a single key pool used in E-G, which substantially increases the key connectivity of the network and improves the network's ability against node captures.

The basic idea of DDHV is as follows. Denote by N the total number of nodes. There are three security parameters used in the protocol, τ, ω, $\lambda(2 \leqslant \tau < \lambda)$. In the key pre-distribution phase, a public $(\lambda+1) \times N$ matrix G (any $\lambda+1$ out of the N columns of G are linearly independent) and ω secret $(\lambda+1) \times (\lambda+1)$ symmetric matrices D_1, D_2, $\cdots$, D_ω are generated over a finite field $GF(q)$, where q is a large enough prime number. Each pair (D_i, G) is called a key space $S_i, i = 1, 2, \cdots, \omega$. For each key space S_i, a symmetric matrix A_i is calculated with $A_i = (D_i \cdot G)^{\mathrm{T}}$. Every node j randomly chooses τ out of the total ω key spaces. For each chosen key space S_i, it saves the j-th-row of A_i ($A_i(j)$) and the seed of the j-th column of G (denoted as $G(j)$) which can be used to generate all the elements in $G(j)$. In the key generation phase, if two adjacent nodes share a key space S_c, they can calculate their session key with the Blom scheme as

$$K_{ij} = K_{ii} = A_c(i) \times G(j) = A_c(j) \times G(i).$$

The DDHV scheme combines the features of the E-G scheme and the Blom scheme; the random key distribution and the session key generation using symmetric matrices. Its security is determined by the parameters τ, ω, and λ. When $\lambda = 0$, the DDHV scheme is the same as the E-G scheme. Compared with the Blom scheme, the key connectivity in DDHV degrades but the network's ability to resist node capture attacks increases. Compared with the E-G scheme, it reduces the probability that different node pairs use same session keys thus improves the security; but it also incurs more storage and computational overhead.

2. *The RS scheme*

The RS scheme proposed by Liu et al.[10] uses random subset assignment in its key pre-distribution phase. It combines the E-G scheme and the Blundo scheme in a similar manner to the combination of E-G and Blom in DDHV. As previously mentioned, the Blundo scheme and the Blom scheme have the same λ-security and key connectivity, thus the performance of RS is similar to that of DDHV. Compared with the E-G scheme, the network's resilience against node captures is improved in the RS scheme. Compared with the Blundo scheme, the key connectivity in the RS scheme is reduced.

The basic idea of RS scheme is as follows. In the key initialization phase, the setup server randomly generates s binary polynomials $f(x, y)$ as the key pool. It then randomly selects t polynomial $f_k(x, y)$ for each node i, and assigns corresponding polynomial shares$f_k(i, y)$ to node i. In the key establishment phase, it uses the same method as in E-G to establish pair-wise session keys or path keys for two nodes in the network.

7.2.4.4 Improving the E-G scheme with nodes' geographic information

In some scenarios, nodes' location information can be used to improve performance of protocols in WSNs. This is because sensor nodes may be unevenly distributed when they are deployed, and nodes located at different positions have different probabilities to be adjacent. For example, if we deploy sensor nodes by dropping them from a helicopter, then nodes dropped at the same place have high probability to be neighbors. If we can acquire the locations of nodes in advance, we can use this information to conduct the pre-distribution of initial keys, which consequently improves performance of key management protocols, e.g., increasing the probability of establishing session keys between neighboring nodes, reducing storage overhead, and increasing the network's ability to resist against node capture attacks.

The schemes that use nodes' location information in key management can be classified into two categories: group-based schemes and grid-based schemes. Group-based key management schemes include CPKS[11], LBKP[11], DR-KPS[12] and DDHV-D[13]; grid-based key management schemes include GKP[10] and PIKE[14]. We point out that, unlike other schemes that will be discussed in the following, the PIKE scheme is not based on the E-G scheme. In PIKE, a node stores pair-wise keys it shares with nodes that reside in the same row or column as itself in the grid. It then uses these nodes as trusted intermediate nodes to generate session keys with other nodes in the network.

1. *The CPKS scheme*

The Closest Pair-wise Key Scheme (CPKS)[11] uses nodes' position information to improve the random pair-wise key scheme[4]. In CPKS, it is assumed that all nodes are deployed in a two dimensional region and each node has a pre-determined deployment location. The main idea of CPKS is to make every node share pair-wise keys with c nodes closest to the node's deployment location. In the key establishment phase, for each node u, the setup server selects a master key K_u for u. For every one of the c nodes closest to u, namely v, the setup server calculates the pair-wise session key between u and v as $K_{u,v} = \mathrm{PRF}_{K_v}(u)$, where PRF is a pseudo-random function. Node u saves all of the c pair-wise keys and node v saves K_v. Node v can calculate $K_{u,v}$ by the equation $K_{u,v} = \mathrm{PRF}_{K_v}(u)$. Using this method, the key generation for newly deployed nodes is simplified.

In CPKS scheme, neighbors of node u only save K_v, which reduces storage overhead. Because every node can obtain its pair-wise keys with its c closes neighbors by looking up in its memory or calculating the keys with *PRF*, the key connectivity in CPKS is high.

2. *The LBKP scheme*

Based on the RS scheme[10], Liu et al.[11] propose a Location-based Key pre-distribution scheme (LBKP) using Bivariate Polynomials. In LBKP, the deployment region is divided into $r \times c$ equal-sized squares. The setup server generates $r \times c$ symmetric bivariate polynomials and assigns each square with a unique bivariate polynomial. For each node p, the setup server preloads five polynomials into p's memory before deployment, including the one assigned to the square within which p resides and the ones assigned to its four adjacent squares. The key establishment phase of LBKP is the same as that of the RS scheme.

The LBKP scheme utilizes nodes' location information to help pre-distribute initial keys, which can effectively improve the network's key connectivity. Meanwhile, the LBKP scheme reserves RS's ability in resisting against node capture attacks. But it incurs more storage and communication cost than RS.

3. *The DDHV-D scheme*

Based on DDHV, Du et al.[13] propose the DDHV-D scheme which aims at reducing the storage overhead of nodes using the knowledge of network deployment. In DDHV-D, nodes are assumed to be deployed in a two dimensional region following a Gaussian distribution. The deployment region is divided into $t \times n$ grids. Nodes are also divided into $t \times n$ deployment groups; the ones in group $G_{i,j}$ are deployed in the corresponding grid. The key space pool S is divided into $t \times n$ sub-pools and each sub pool $S_{i,j}$ is related to a group $G_{i,j}$. If two grids are adjacent, there are common keys between the two key space pools related to node groups deployed in them. If two grids are not adjacent, there are no common keys between the corresponding key space pools. The key pre-distribution phase and the key establishment phase are similar to that of DDHV; the only difference is that when choosing τ key spaces for a node, the key spaces are not chosen from the total key space pool S but instead from a sub key space pool related to the group the node belongs to.

In the DDHV-D scheme, nodes deployed in adjacent grids have high probability to have shared keys, thus have high probability to establish their session keys. Compared with DDHV, DDHV-D achieves the same key connectivity with less storage overhead.

4. *The GKP scheme*

Liu et al. propose a Grid-based Key Pre-distribution scheme (GKP) which combines the E-G scheme and the Blundo scheme in a manner different from the RS scheme[10]. The authors propose to divide the deployment region into $m \times m$ grids; nodes are assumed to be deployed at grid points. In the key pre-distribution phase, the setup server generates 2m bivariate polynomials $\{f_i^c(x, y), f_i^r(x, y)\}, i = 0, \cdots, m-1$, and assigns the node at grid point (i, j) with a pair of polynomial shares $f_i^c(x, y)$ and $f_j^r(x, y)$. In the shared key discovering phase, if two nodes are in the same row or column, they generate their session key using the Blundo scheme; otherwise they try to establish their session key using the path key establishment method proposed in E-G scheme.

The GKP employs the Blundo scheme to establish pair-wise session keys, thus it is λ-secure. It utilizes the location information of nodes to pre-distribute initial keys, which provides the same key connectivity and security as do the Blundo scheme with less storage and computational overhead.

7.2.5 Hierarchical key management schemes

WSNs may be heterogeneous or dynamic. There are usually two types of nodes in a wireless sensor network, one with limited resources to be used for data collection and one with strong abilities which can be used for some management tasks. Sensor networks composed of nodes with different abilities are heterogeneous. Sensor networks may also exhibit dynamic properties. For example, the topology or the connectedness of the network may be changed due to node failures.

Taking these characteristics into account, in recent years some hierarchical key management schemes have been proposed, which are considered as a tradeoff between centralized schemes and purely distributed schemes. In hierarchical key management schemes, nodes in the network are grouped into clusters. In each cluster there is a cluster head with strong ability to perform key distribution, generation, or update for all nodes in that cluster.

This type of key management scheme reserves the advantages of centralized schemes, e.g., low storage and computational overhead, meanwhile weakens the dependence on the base station to manage keys for all nodes and improves scalability.

Typical hierarchical key management schemes include the Localized Encryption and Authentication Protocol (LEAP) proposed by Zhu et al.[5], the Unbalanced Random Key Pre-deployment (URKP) proposed by Traynor et al.[15] that considers heterogeneity of sensor nodes and dynamic topology in real deployed WSNs, the SHELL protocol[16], the LOCK protocol[6] and the EEHS protocol[17] that all adopt the EBS mechanism[18] cluster-based hierarchical framework to perform dynamic key management, and the Asymmetric Key Pre-distribution Scheme (AKPS)[19] proposed by Liu et al. These protocols reserve some advantages of centralized key management schemes, e.g., low computational and storage overhead. Compared with centralized schemes, these protocols rely less on the base station, thus achieve high scalability as do distributed key management schemes. In the following, we use the LOCK scheme and the LEAP scheme as examples to explain the basic ideas of hierarchical key management schemes.

1. *The LEAP scheme*

Zhu et al.[5] propose the LEAP protocol which aims at supporting in-network processing and restricts the threat of a compromised node into a small neighboring region of the node. The LEAP scheme uses multiple kinds of keys to provide diverse security level for different type of messages. There are four types of keys generated for a node in LEAP: an individual key shared between the node and the base station, a pair-wise key shared only between the node and one of its direct neighbors, a cluster key shared by nodes in the same cluster, and an identical group key shared by all nodes.

The procedure to generate these keys is as follows. Before the nodes are deployed, the setup server randomly selects a master key K_s and generates an individual key for every node using this master key and a pseudo-random function. Every node stores its individual key before deployment. In the key pre-distribution phase, the setup server generates an initial key K_I and stores it in every node; any node u uses this initial key and a pseudo-random function to generate a master key K_u.

In the key establishment phase, a node u first sets a timer and broadcasts a HELLO message to its neighbors. Upon receiving the replied ID from a neighbor node v, node u calculates node v's master key K_v using K_I and node v's ID. It then authenticates node v with K_v and its pair-wise key shared with v by $K_{uv} = f_{K_v}(u) = f_{K_u}(v)$. When the timer expires, node u erases the initial key K_I and all the master keys of its neighbors but keeps its own master key K_u.

In the cluster key establishment phase, the cluster head randomly generates a key as the cluster key and sends this key to its cluster members. The key sent to node v is encrypted using the pair-wise key shared between v and the cluster head; thus only node v can decrypt it. When cluster members are revoked, the cluster head generates a new cluster key and updates it to all cluster members in the same way.

The group key, which is shared by all the nodes in the network and the base station mainly used to encrypt broadcasting messages sent by the base station. When establishing or updating the group key, the base station first encrypts the group key with its cluster key and broadcasts the encrypted key to its children nodes in the same cluster. The children nodes decrypt the group key and relay to their own children nodes in a same manner. This procedure is executed iteratively until all the nodes in the network obtain the group key. In this process, the μTESLA protocol is used to authenticate messages sent by the base station, which prevents an outsider adversary or a compromised node from impersonating the base station.

In the LEAP scheme, the establishment and update of pair-wise keys are carried out in a cluster, restricting the threat of a compromised node into an immediate neighborhood of that compromised node. The disadvantage is that a network-wide initial key must be retained for a period of time after the deployment of the network; if this key is exposed during this time, the entire network will be threatened.

2. *The LOCK scheme*

Based on the EBS mechanism, Eltoweissy et al.[6] proposed the Localized Combinatorial Keying (LOCK) scheme. EBS is a combinatorial optimization method that can be used in key management protocols. It is usually expressed as EBS(n, k, m), where n is the number of nodes, k is the number of keys to be managed for each node, and m is the number of messages to be broadcasted when a node updates its administrative keys. Two types of keys are used in EBS-based key management schemes: administrative keys and session keys. Administrative keys are used to generate initial keys, generate or update session keys, or revoke keys of compromised nodes.

LOCK uses a three-tier network structure: the base station is the first tier, all cluster heads form the second tier, and all other nodes form the third tier. All cluster heads in the second tier form a group called the cluster head group. LOCK uses two levels of administrative keys: the first is used to generate and update group session keys used in the communications between the base station and the cluster head group, and the second is used to generate cluster session keys used in the communications between a cluster head and its cluster members.

In the initialization phase of LOCK, each sensor node establishes a set of backup keys only shared between itself and the base station. These keys are used to authenticate newly deployed cluster heads. Because cluster heads do not know the backup keys, LOCK achieves good resistance against node captures. Furthermore, because of its clustered structure, LOCK has good scalability and achieves high security by limiting the impact of a compromised node into a local part of the network.

7.2.6 Future research directions

Although there have been many efforts dedicated to key management in WSNs, there are still some issues unsolved. From our point of view, we list some potential future research directions below.

1. *Supporting more communication types*

Most existing key management schemes only consider how to establish pair-wise keys between neighboring nodes which can support unicast (point-to-point) communication. However, many messages in WSNs need to be broad-casted or multi-casted to a set of sensor nodes. Key management schemes that can provide different session keys to support more communication types, need more focus.

2. *Dynamic key management*

Some nodes in the network may be compromised by adversaries. When this happens, the compromised nodes should be excluded from the network and keys related to them should be revoked and updated dynamically. Most existing key management schemes do not provide dynamic key management or perform dynamic key management in centralized manners. Centralized schemes usually incur high communication and computational overhead. Thus we need to design schemes in which dynamic key management is performed via collaboration among nodes to provide good scalability and to reduce computational and communication overhead.

3. *More effective authentication mechanisms*

Both source node authentication and message authentication are necessary to provide a guarantee of security when generating session keys via negotiation among nodes. However, the message authentication code (MAC) mechanism is vulnerable (MAC can be faked), while the digital signature mechanism based on asymmetric key mechanisms is not suitable for WSNs. It is an important research issue to design light-weight authentication mechanisms that can provide enough security and are suitable for WSNs.

7.3 Secure routing protocols in WSNs

Many routing protocols designed for WSNs pay little attention to security issues. In this section, we will first discuss typical security threats that routing protocols face in WSNs and general strategies to defend them, then survey typical secure routing protocols. We also suggest potential research directions at the end of this section.

7.3.1 Typical attacks and general defending strategies

7.3.1.1 Typical attacks to routing protocols

Attacks launched at the network layer can be classified into two categories according to their targets[20]. The first category attempts to access or directly manipulate user data, e.g., selective forwarding, Sybil attacks, acknowledgement spoofing and passive eavesdropping. The second category attempts to affect the network's routing topological structure, such as spoofed routing information, Sinkhole attacks, Wormhole attacks, and HELLO flooding attacks. We briefly describe them in the following.

1. *Selective forwarding*

In this type of attack, malicious nodes selectively forward or refuse to forward received packets to make them fail to reach their destinations. In order to reduce the possibility of their illegal behaviors being detected, malicious nodes may only discard or alter packets from targeting nodes, while forwarding packets from other nodes normally.

When the attacker is on the data transmission path, selective forwarding is most effective. If the target data flows do not pass the attacker but pass its neighboring nodes, the attacker can jam the transmission of target packets or produce collisions on the transmitting channels to ruin the target data packets, which in fact implements a selective forwarding attack successfully[21–23].

2. *Sybil attack*

In Sybil attacks, a malicious node behaves like many legitimate nodes by faking multiple legal node IDs. It can then modify, selectively discard or forge packets. It can also eavesdrop on passing data flow. There are two types of Sybil attacks[24,25]: in the first type the malicious node forges several legal IDs in one location, while in the second type the malicious node forges multiple IDs at diverse locations.

Sybil attacks are very typical in WSNs. If combined with other attacks, Sybil attacks can cause great harm to WSNs[26]. For example, Sybil attacks can cause serious damage to geographical routing protocols by faking multiple legitimate nodes at different locations. It can also degrade the performance of location-based redundancy schemes.

3. *Acknowledgement Spoofing*

Malicious nodes eavesdrop on packets addressed to their neighbors and forge acknowledgements to overheard packets. This can result packets to be transmitted on communication links with low quality or delivered to fake nodes. Acknowledgement spoofing can cause packet loss, and can be used to launch selective forwarding attacks[22,23,27].

4. *Passive eavesdropping*

The attacker overhears the information on links and extracts the traffic pattern by analyzing the eavesdropped data. It can then deduce some sensitive information of the overheard node based on which it can launch the most effective attacks.

5. *Spoofed routing information*

Attackers can spoof, alter or replay routing information to generate false routing information and create routing loops, extending or shortening source routes.

A spoofed routing information attack can cause direct damages on routing protocols because it uses routing information exchanged among nodes. It may make the network partitioned, cause congestions or enlarge end-to-end packet delivery latency.

6. *Sinkhole attack*[24]

In this type of attack, malicious nodes mislead their neighboring nodes to select themselves or other compromised nodes as relaying nodes in their routes, resulting in sinkholes around malicious nodes which pull data packets and prevent these packets from reaching their original destinations.

Sinkholes can attract almost all the data flow in specific areas, preventing corresponding packets from reaching their true destinations. Furthermore, the dupe nodes may spread the information of the sinkhole, thus extending its operation range, making the case worse because more data flows will be attracted by the sinkhole[28–30]. Meanwhile, the adversary may alter, selectively discard, forge, or eavesdrop on all packets passing the sinkhole, which makes it convenient for the adversary to combine sinkhole attack with other attacks.

7. *Wormhole attack*

In a wormhole attack, two malicious nodes are connected with a direct low latency link called wormhole link. With the wormhole link, the adversary can capture data transmissions on one node, send them quickly to the other node through the wormhole link and replay these data transmissions. Current solutions on wormhole attacks mostly rely on fine-grained time synchronization or precision position information of nodes. In WSNs, wormhole attacks are difficult to be detected because it is hard to get this information with resource-constrained sensor nodes.

8. *HELLO flooding*

In this type of attack, malicious nodes broadcast HELLO packets to its neighboring nodes and convince them to establish routes passing them. The goal of HELLO flooding attacks is to make the network into a chaos state, preventing legitimate data packets from reach their destinations[24,27]. To achieve this, the adversary only needs to broadcast its HELLO messages with large enough power. Because many routing protocols rely on local HELLO messages exchanged between neighboring nodes, they are vulnerable to HELLO flooding attacks.

Among the aforementioned attacks, Sybil attack, Sinkhole attack and Wormhole attack are most basic attacks[20]. They are highly destructive to WSNs because they can alter, discard, forge or eavesdrop on data packets. They are usually combined with other types of attacks when the adversary launches routing layer attacks. In recent years, many researchers carried out detailed analysis on these attacks, especially Sybil, Sinkhole and Wormhole attacks. A number of general strategies to defend these attacks are proposed; we briefly describe them in the following.

7.3.1.2 Classification and vulnerability of routing protocols in WSNs

Generally, routing protocols in WSNs can be classified into five categories: TinyOS beaconing routing, data-centric routing, clustering-based routing, location-based routing and energy-aware routing[24,31–33].

1. *TinyOS beaconing*

In this type of routing protocols, each node has a unique address. The sink node periodically broadcasts messages indicating a route update. Upon receiving the update message, a node set its parent node as the node from which it receives the update message and rebroadcasts the update message to other nodes. In this way, a breadth-first spanning tree rooting at the sink node is constructed which acts as the routing tree.

TinyOS beaconing[24] is relatively simple and does not have any safe measures during the route update process, so it is vulnerable to malicious attacks. Attackers can launch Wormhole attacks or Sybil attacks to lead the data flow to pass through the malicious node. They can also launch spoofed routing information to form routing loops, or launch HELLO flood attacks to make the network chaotic. In addition, if the malicious nodes are on the data transmission path, they can selectively forward data packets thus damage the data transmission directly.

2. *Data-centric routing*

These protocols describe data using property-based naming schemes. The sink node sends query requests to a specific region to get routing information; the data is transmitted in the reverse direction of the query path and may be aggregated to save energy consumption. Typical data-centric routing protocols include Directed Diffusion (DD)[34], SPINS[2], and Rumor[35].

In these protocols, the base station sends requests to nodes by flooding. Nodes then send the data to the base station on the reverse path. Therefore, when the malicious node forges a request, it can easily eavesdrop on the data, mislead the data transmission path, launch selective forwarding attacks. In addition, data-centric routing protocols are vulnerable to Wormhole attacks and Sybil attacks.

3. *Cluster-based routing*

In cluster-based routing protocols, the entire network is divided into several clusters and each cluster has a cluster head that is in charge of collecting data from cluster members and sending the collected data to the sink node, with optimal data fusion on the cluster head in order to reduce transmitted data volume. Typical clustering-based protocols include LEACH[36], TEEN[37], and PEGASIS[38].

In cluster-based protocols, nodes choose the cluster head with the highest received signal strength indicator (RSSI) and join that cluster. Thus the adversary can launch HELLO flooding attack to make a large number of nodes join the cluster in which it is the cluster head. The attacker can further launch selective forwarding attack or tamper data to further damage the function of the network. Furthermore, the adversary can launch Sybil attacks to increase its possibility of being elected as cluster head, even if the cluster head is randomly selected and is different in different rounds.

4. *Geographical routing*

In this type of routing protocols, every node is assumed to be aware of its physical position and also knows the position of its destination node. When forwarding data packets to the destination node, greedy strategies are used, e.g., the node selects from its neighbors the closest node to the destination or the farthest node from the current node as next hop relaying node. Typical geographical routing protocols include GEAR[39] and GPSR[40].

Since geographical routing protocol nodes are assumed to be aware of their locations, they are vulnerable to acknowledge spoofing attacks. An attacker can report a false location to increase its probability of being on a target data transmission path. In addition, the malicious node can launch Sybil attacks to forge identities of multiple locations in order to increase its chances in placing itself on the path of any nearby data flow; afterwards it can further launch selective forwarding attacks. As GEAR always assigns routing tasks according to nodes' residual energy, the attacker can always claim to have the highest residual energy. In GPSR, a malicious node may make a false location statement to construct routing loops, which will disrupt normal data transmissions.

5. *Energy-aware routing*

When WSNs are deployed in adverse environments, energy saving must be considered. According to the distribution of remaining energy in different areas, energy-aware routing protocols establish the optimal path in terms of energy consumption or the path that can achieve the longest network lifetime. In energy aware routing protocols, a malicious node can use a high-energy machine to launch Sybil attacks and HELLO flood attacks. Typical energy-aware routing protocols include SPAN[41] and GAF[42].

7.3.1.3 General defending strategies

In order to prevent external attacks in WSNs, a general method is to use encryption and authentication on the link layer. We can encrypt data packets transmitted on wireless links, or authenticate the identity of the source node or the destination node. These strategies can effectively resist most external attacks, including passive eavesdropping, external Sybil attacks, acknowledge spoofing and HELLO flood attacks.

For internal attacks such as Sybil attacks, wormhole attacks and sinkhole attacks, the following strategies are proposed.

1. *Encryption and authentication*

With encryption and authentication, nodes can authenticate identities of each other and prevent malicious nodes from joining the forwarding path. Encryption and authentication need to distribute keys among nodes in a WSN; distribution and management of keys in WSNs is described in Section 7.2. Strategies based on encryption and authentication cannot prevent comprised nodes that have legitimate keys from joining in the forwarding path. Moreover, such schemes incur high computational overhead, which limits their applications in securing routing protocols in WSNs.

2. *Multi-path routing*

Nodes can dynamically select next hop relaying node when forwarding data packets, which establishes multiple paths to the destination node. This strategy, termed multi-path routing in this chapter, can effectively reduce the opportunity that malicious nodes obtain complete control on the target data flow.

Zhang et al.[43] propose a novel safe anonymous multi-path routing strategy that makes it difficult for the adversary to discover the key nodes between the source and the destination by traffic analysis and hence cannot launch wormhole attacks. This is achieved by using anonymous identity and hiding location of the communication nodes.

Wang et al.[20] propose a malicious node detection and localization strategy by combing multi-path routing and source coding. In this strategy, the source node first encodes the data such that the encoded data can be used to detect malicious nodes. The encoded data is then sent to the destination via multiple established paths. Upon receiving the encoded data, the destination node extracts the corresponding information in order to detect potential malicious nodes on the transmission paths. If malicious nodes are detected, the result will be announced to intermediate nodes on which the malicious nodes exist and the malicious nodes will be isolated. Theoretical analysis and simulation results both show that this strategy can effectively locate the malicious nodes therefore can defend against wormhole attacks and Sybil attacks effectively.

Because data reach the destination along different paths in multi-path routing strategies, this type of strategies can effectively defend selective forwarding, sinkhole attacks, wormhole attacks and Sybil attacks. However, multi-path routing strategies need some time to establish acyclic multiple paths which inevitably increases deliver delay. Furthermore, each node needs to maintain a routing table for each path and thus the size of the routing table is proportional to the number of existing paths, which increases the maintaining overhead of routing tables.

3. *Location-based detection strategy*

The goal of this type of strategy is to prevent malicious nodes from occupying the path by using wormhole attacks. In wormhole attacks, the distance claimed by malicious nodes is shorter than the actual distance; thus malicious nodes can be detected by comparing the estimated distance and the distance claimed by the malicious nodes.

Hu et al.[44] propose a method that uses geographic-constrained and time-constrained packets to detect wormhole attacks. It assumes that fine-grained clock synchronization can be provided by special hardware like GPS and set the maximum transmission distance and the maximum survival time for the transmitted packets. Therefore, if the target node detects that the transmission time or transmission distance of the received packets exceed corresponding threshold, it knows there are wormholes. Wang et al.[45] propose EDWA, a method that assumes nodes are aware of their positions and the distance between two nodes in terms of hop count can be calculated. If the calculated hop count is larger than the hop count in the acknowledge packet, it is considered that wormholes exist in the network. In addition, the strategy can locate the malicious node in a small region. Hu and Evans[46] propose a method to establish reliable neighborhood relationship between nodes by using directional antennas. Each node checks the source direction of the received signal; only if the directions of the two sides match, the neighborhood is confirmed. In reference [47] the authors propose a method to discover malicious nodes by detecting the bending properties of the reconstructed network topology plane using intermediate controllers.

Location-based detection strategies need support of GPS or similar hardware devices, which not only increases the overhead but also limits their application in WSNs.

4. *Strategy based on monitoring and reputation management*

This type of strategies determines whether the packets are altered by eavesdropping packets forwarded by neighbors or assigns different credibility to nodes. When choosing next hop relaying node to forward the packet, a node selects those nodes with large credibility value to avoid malicious nodes.

Issa Khalil et al. present the LiteWorp protocol[48] that monitors and records the forwarding and transmission of data packets to detect malicious nodes. When the malicious behavior record of a node exceeds a threshold, that node will be determined as a malicious node and removed from the network. Liang and Fan[49] propose to assign credit levels to neighboring nodes by eavesdropping on their transmitted data and choose forwarding path based on the credit levels. Strategies based on monitoring and reputation management require a large number of nodes to be involved in monitoring for a long time, which consumes a lot of energy. If the nodes run out of energy prematurely, the network will be paralyzed.

7.3.2 A typical secure routing protocols in WSNs: INSENS

Deng et al.[50] proposed an intrusion-tolerant routing protocol for WSNs (INSENS) that aims at defending sinkhole attacks, flooding attacks, and spoofed routing information attacks. It provides an approach to construct secure tree-based routing structures by employing one-way hash, symmetric encryption and authentication.

Before the deployment of a WSN, the base station first generates a sequence $n_1, n_2, \cdots, n_k$ using a one-way hash function F, where n_1 is a random number and all n_i satisfy $F(n_i) = n_{i+1}$. Each sensor is assigned with a generated number n_k. Every sensor knows the hash function F and has a pre-distributed key shared with the base station.

The secure routing discovery phase is as follows. The base station broadcasts a routing request in the format of $\{$*type*, *OWS*, *size*, *path*, *MACR*$\}$ to collect the topology information of the network. In order to defend against the replay attacks and provide identity authentication, in the ith routing request the base station sets OWS_i as n_{k-i}, and the node that receives the ith request can verify if the request is sent by the base station by calculating F^i (OWS_i) and comparing the result with n_k. Because F is a one-way hash function, the malicious node cannot infer n_j with n_i when $j < i$, thus cannot impersonate the base station to broadcast routing requests. Every sensor node saves the newest *OWS* as OWS_{fresh}. When a sensor node receives a request whose *OWS* value is older than OWS_{fresh}, it judges the request as a duplicate and discards this request. With this mechanism, INSENS can defend flooding attacks.

In INSENS, before forwarding a routing request, an intermediate node first marks the node from which it receives the request as its parent node. It then adds itself into the path and update the value of *MACR* in the request as $MACR{=}MAC(size|path|OWS|type,Key)$ and forwards this request to other sensor nodes. Meanwhile, it records the old *MACR* as *parent_info*. In this protocol, every node needs to report its connectivity topology information to the base station. An intermediate node x receiving the routing request sends a feedback packet to the base station in the reverse path of the routing packet. The feedback packet has the format of $\{$*type*, *OWS*, *parent_info*, *path_info*, *nbr_info*, *MACF*$\}$ where $MACF{=}MAC(path\_info|nbr\_info|OWS|type, Key)$. The base station can use *MACF* to check if the feedback packet is sent by the node x and if the content of the packet is altered in transmission. When receiving a feedback packet, the base station uses the *parent_info* (recording the *MACR* of x's parent node), the *path_info* (recording the path from the base station to node x and x's *MACR*), and the *nbr_info* (recording all x's neighbors' *MACR*) to construct local topology of x. If x is a malicious node, the base station detects this by observing inconsistency between x's feedback information and its neighbors' feedback information. Then the base station calculates the forwarding table for each sensor node and constructs a tree-based routing structure rooted at the base station. The routing tree is sent to all sensor nodes in a breadth-first manner; and data is sent to the base station in multi-hop manners.

There are other secure routing protocols for WSNs. For example, the Feedback towards dynamic Behavior and Secure Routing (FBSR) proposed in reference [51] is a security routing protocol based on feedback. It employs a trajectory tracking mechanism to detect malicious behaviors of attackers, and isolates the malicious node from the data delivery path in order to defend attacks launched by the malicious node. The SLEACH protocol proposed in reference [52] is an improvement over the LEACH protocol. It uses authentication and a reputation mechanism to defend selective forwarding attacks.

7.3.3 Future research directions

Due to the characteristics of WSNs, we think the following are potential research directions in the future.

1. *Secure localization technology*

With accurate location information of nodes, the base station can easily detect malicious nodes that try to fake identities at false positions. Combining this technology with other security mechanisms, this technology can be used to defend attacks such as wormhole attacks and Sybil attacks.

2. *Dealing with capture attacks*

Current routing protocols are usually vulnerable to node capture attacks; the disclosure of a single node may ruin the functionality of the entire network. It is important to design routing protocols that can resist node capture attacks. Authentication mechanism may be useful to prevent malicious nodes from denying their previous behaviors and to isolate malicious nodes from the data transmission path.

3. *Path hidden technology*

Passive attacks cause great security threats to WSNs; path hidden technology can be used to prevent passive attacker from detecting the network topology and key node thus fundamentally enhances the security of routing protocols. In order to achieve path hidden, we can use the fake identity mechanism (namely node use fake identity instead of their real identity to communicate and change the fake identity regularly or irregularly) or use onion routing technology to hide the path in onion hierarchy.

7.4 Location privacy protections in WSNs

In some WSNs applications, the exposure of some key nodes' location information will cause severe negative results to the network. For example, when WSNs are used in battlefields for communication, location information of soldiers or headquarters is extremely sensitive. The sensor nodes carried by soldiers or monitoring their activities should not expose the location privacy of the soldiers in the communication process. Meanwhile, headquarters should not expose their location privacies when they are sending commands or receiving reports. Similarly, when WSNs are used to monitor wild animals, the locations of wild animals are also extremely sensitive. Sensors monitoring activities of wild animals should not expose the locations of monitored animals when collecting related data.

The goal of location privacy protection is to prevent some key nodes' locations in WSNs from being exposed. Existing location privacy protection protocols can be divided into two categories according to their protecting targets: those who try to protect location privacy of source nodes and those who try to protect location privacy of the sink node.

1. *Source node location privacy protection*

When WSNs are used for monitoring precious resources such as wildlife animals, sensors that monitor the protected objects usually act as source nodes are the direct source from which the information about protected objects is obtained. By tracing source nodes, an adversary can easily find the protected objects and expose their location privacies. Thus it is important to protect the location of source nodes in such applications. Many source node location protection protocols have been proposed, including the Phantom Routing protocol[53], the source-location privacy protocol based on locational angles[54], the Cyclic Entrapment Method protocol[55], the Greedy Random Walk protocol[56], and the Self-adjusting random walk protocol[57], etc.

2. *Sink node location privacy protection*

The sink node connects the sensor network with external networks. All the data collected in the network should be transferred to the sink node first before they can be accessed by external users. Furthermore, the sink node usually plays the role of an administrator of the entire network. Once it is compromised, the security of the whole network will be threatened. Thus the location privacy of the sink node is extremely important in the network and should be well protected. Existing sink node location privacy protection protocols include the Decoy Sink Protocol[58], the Location Protection Route[59], the Differential Enforced Fractal Propagation[60], etc.

7.4.1 Attack models

There are two types of attacks that may threaten location privacies of nodes in a WSN: internal attacks and external attacks. In internal attacks, the adversary has exact knowledge of the formats of the packets exchanged in the network and can extract their contents based on this. In external attacks, the adversary has to infer the states of the network by observing its data flows. The wide application of encryption of communication links (see Section 7.2) makes it difficult for the adversary to launch internal attacks; thus current researches on location privacy mainly focus on external attacks. We introduce three typical external attacks in the following.

7.4.1.1 Attacks tracing source node locations

This attack model assumes that the attackers are equipped with devices that can monitor or locate wireless signals, with which the attackers can monitor the behavior of data transmissions within a certain area. It is assumed that the ability of the attackers is nearly the same as normal nodes, thus they usually can only monitor data transmissions in one-hop range. The attackers track in the opposite direction of data packet transmissions when they try to trace the source node. A typical scenario is shown in Fig. 7.1. In this scenario, an attacker first stays at the sink node waiting for reported data packets. When it detects the arrival of data packet m_1, it can infer the location of the sender of m_1, in this case B, with its wireless signals locating device. Then it moves to B and repeat this procedure. As long as the source node sends enough packets to the sink node, the attacker can always successfully trace the location of the source node in this hop-by-hop manner.

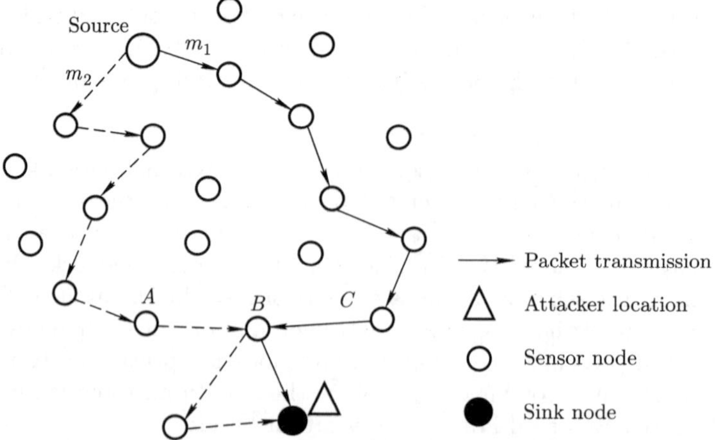

Fig. 7.1 Trace the source node hop-by-hop.

In this attack model, after the attacker has traced to an intermediate node, it stays at that node and waits for following data packets to continue the trace procedure. Some variants of this attack model assume the attacker can look backward, i.e., if the attacker detects no packets for a long time, it will move back to the previous traced node and restart the tracing procedure. For example, as shown in Fig. 7.1, when the attacker traces to the node A, the route between the source node and the sink node changes from the dashed line to the solid line so that the attacker will not hear any packets sent from the source node. In this case, the attacker may move back to node B, restart the monitoring procedure at B, and finally successfully traces to the source node along the new route.

7.4.1.2 Attacks tracing the sink node location

This attack model also assumes that the attackers are equipped with wireless signal monitoring and locating devices. According to the time stamps of received data packets, the attacker determines which nodes are on the transmission path and move to the sink node in the reverse path. The tracking process is shown in Fig. 7.2. Initially, the attacker stays at node A and monitors the passing data packets in its one-hop range. If it hears that node B always resends the packet that node A sends, the attacker may infer that the packets are transmitted along a path from node A to node B. It then moves to node B and repeats this procedure until it reaches the sink node.

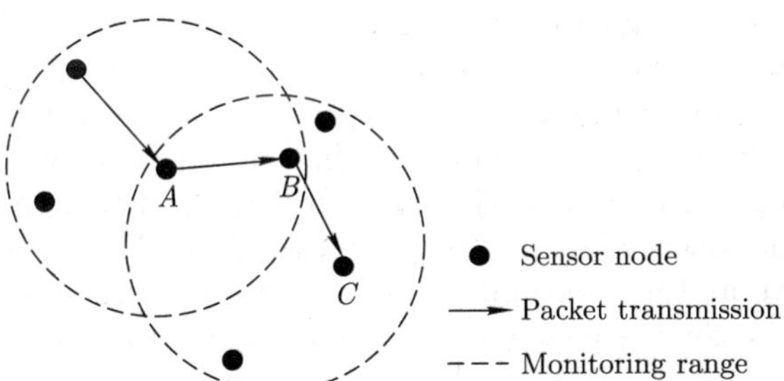

Fig. 7.2 Trace the location of the sink node.

7.4.1.3 Attacks based on traffic analysis

This attacker model assumes that the attacker can monitor the traffic in the network, i.e., it can monitor wireless communication traffics of different parts of the network or the total traffic of the entire network for a period of time. For example, the attacker can deploy a large number of low-cost devices to overhear the global traffic of the network. By analyzing the traffic patterns, the attacker can infer the location of the source node or the sink node. In reference [60] the authors have studied network traffic patterns when shortest paths are used to routing and forwarding packets. Because there are less nodes that can play the role of forwarders near the sink node, average traffic load of nodes near the sink node are significantly higher than other nodes in the network. The attacker can infer the location of the sink node by comparing different nodes' traffic loads.

7.4.2 General location privacy protection strategies

Currently, a lot of defense strategies have been proposed to protect the location privacies of key nodes in a WSN from being exposed. They can be roughly divided into four categories. We elaborate them in the following.

7.4.2.1 Flooding

Ozturk et al. proposed the first source node location privacy protection protocol using flooding for WSNs[61]. They used a metric called safety period to evaluate the performance of a location privacy protocol in the presence of a local attacker. The safety period is defined as the number of messages the source node can send before it is localized by the attacker. With this metric, they have evaluated the impacts of three flooding mechanisms on the privacy of source node locations: baseline flooding, probabilistic flooding, and phantom flooding.

1. *Baseline flooding*

In this flooding mechanism, every sensor node checks whether a received packet is duplicated. It rebroadcasts the packet to all neighbors if it is not, otherwise it discards the duplicated messages. Because all nodes participate in the flooding process, it was believed that the attacker will be effectively misled to wrong source nodes. However, in practice the attacker can easily trace to the true source node in this type of flooding. This is because the first packet to arrive at the sink node is in fact transmitted along the shortest path between the source node and the sink node; thus the attacker can easily trace the true source node reversely along this shortest path.

2. *Probabilistic flooding*

To address the side effects of baseline flooding, probabilistic flooding is proposed in reference [61], in which intermediate sensor nodes forward packets in a probabilistic way. Upon receiving a packet, a sensor node uses a predetermined probability to determine if it should forward the packet. With this method, the route used to deliver the packets from the source node to the sink node are not fixed, which makes it more difficult for the attacker to trace the source node. Nonetheless, it is not guaranteed that all data packets sent by the source node will be received by the base station due to the randomness involved in this approach.

3. *Phantom flooding*

In phantom flooding, it takes two steps to deliver a packet from the source node to the base station. In the first step, the packet is sent to a random node called phantom node by random walking or direct walking. In the second step, the packet is flooded by the phantom node into the network to reach the base station. The randomness involved in the first step increases the difficulty for the attacker to trace the source node, thus prolongs the safety period. However, with phantom flooding the transmission latency of packets also increases.

Although flooding strategies can help protect the source node location privacy, it is still relatively vulnerable to the hop-by-hop tracing attacks. Furthermore, flooding will consume a large amount of energy in the network and hence may substantially reduce the lifetime of the network.

7.4.2.2 Random walk strategies

The basic idea of random walk strategies is that every packet takes a different route to the sink node. For every packet sent by the source node, the transmission path is randomly generated therefore not fixed, which increases the length of data transmission paths and decreases the number of packets passing an individual node. With this type of strategy, the attacker may not be able to obtain enough packets to trace the source node successfully. Typical random walk based strategies are described in the following.

1. *Phantom routing protocol*

Phantom routing is proposed to protect the source location. In the phantom routing protocol, data packets are forwarded randomly for several hops using the random walk mechanism. Therefore, it is difficult for the external attackers to trace back and locate the source location. A typical scenario is shown in Fig. 7.3.

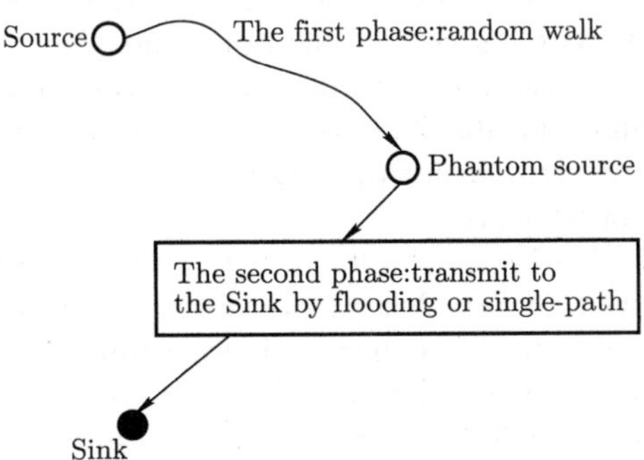

Fig. 7.3 The two phases of phantom routing.

As shown in Fig. 7.3, phantom routing is a two phase routing protocol. In the first phase, the source node randomly forwards the data packets to a random node called phantom source using the random walk mechanism. In this phase, the source node may forward packets completely randomly or randomly in a given direction, which will make the phantom source far from the real source. In the second phase, the phantom source floods the packets into the whole network or transmits the packet using a single path to reach the sink node.

As mentioned earlier, the phantom flooding routing scheme may result in high energy consumption. When using flooding in the second phase, it may make the attacker more likely to capture packets and trace the source node faster. On the other hand, if single path routing is used in the second phase, phantom routing can save energy greatly and makes it more difficult for the attacker to trace the source node successfully.

2. *Locational angle-based phantom routing*

The aforementioned phantom routing protocol with single path can balance safety period and energy cost well. However, it uses a pure random walk mechanism to choose the phantom source, which usually enlarges the length of data transmission path which makes the improvement on safety period insignificant.

In reference [54] the authors proposed a locational angle based phantom routing protocol which improved safety period by reducing "wasting paths". In the proposed protocol, a node selects its relaying node based on a probability determined by the angle at a neighboring node formed by two line segments connecting the source node, the neighboring node and the sink node. The basic idea is to select nodes with larger angles in order to reduce wasting paths and prolong the safety time.

3. *Location Protection Route mechanism* (*LPR*)

LPR[59] is proposed to protect the sink location privacy. In this strategy, the attacker model tracing to the sink hop-by-hop is first characterized. The attacker first infers the direction of packet routing by monitoring temporal correlation between wireless communications and then moves towards the sink node in this direction. By tracing packet transmissions continuously, the attacker can finally locate the sink node's location. In LPR, the authors proposed to combine random forwarding and the packet-faking mechanism to defend hop-by-hop attacks.

Each sensor divides its neighbors into two lists: a closer neighbor list containing neighbors that are closer to receiver, and a farther neighbor list containing other neighbors. After the two lists are built, LPR works as follows. When a sensor tries to forward a packet, it will select the next hop node from the further neighbor list with probability P_f and select from the closer neighbor list with probability $1 - P_f$, where P_f is a system parameter. By adjusting the value of P_f, one can tune the tradeoff between energy efficiency and location privacy.

In LPR, the next hop from a sensor to the receiver is unfixed. Sometimes the next hop is even farther away from receiver, which makes it harder for the adversary to successfully launch packet-tracing attacks. As long as $P_f < 50\%$, LPR can guarantee that every packet will be delivered to the receiver. It is easy to implement and only requires one packet broadcasted from the receiver (every time it moves to a new position) to setup the routing structure. It allows the network designer to make flexible tradeoff between energy efficiency and protection strength through tuning a system parameter.

The adversary can still expose the location privacy of the sink node by analyzing overall traffic trends in the network in the LPR protocol. A higher value of P_f can alleviate this problem, resulting in longer packet delivery delay and more energy consumption. Furthermore, the attacker can stay at one location and keep eavesdropping for a certain period of time. To guarantee packets can be delivered to the receiver eventually, P_f must be smaller than 50%, which means that sensor node is more likely to forward packets to nodes in the closer neighbor list. Thus most packets flow from a sensor to the receiver. If the attacker overhears enough large number of packets, it can figure out the direction of the packet flow and search for the receiver along this direction.

To address this problem, an additional mechanism is introduced to smooth the traffic trend in the network by sending fake packets in the direction away from the receiver. In combination with the fake packets mechanism, the LPR protocol effectively prolongs the safety time of key nodes in the network.

4. *Differential Enforced Fractal Propagation* (*DEFP*)

DEFP adopts several correlation eliminating mechanisms to prevent adversaries from exposing sink location privacy via traffic relation analysis. Similar to the LPR protocol, DEFP also uses random forwarding and packets faking to eliminate a smooth traffic trend. In this scheme, each node has multiple parent nodes which route messages to the base station. When forwarding a message, a node randomly selects one of its parent nodes as the next-hop node. This scheme can be enhanced using controlled random walk. When forwarding a message, it selects one of its parent nodes as next hop node with probability p, and selects one from its neighbors with probability $1-p$. This technique introduces additional delivery time delays, which are proportional to extra hops taken by the messages to reach their destination.

In this protocol the authors propose to generate differential numbers of fake packets for different nodes. Nodes experiencing light traffic generate large number of fake packets, while nodes experiencing heavy traffic generate less or none. With this mechanism, the traffic trend is smoothed. Fake packets are randomly forwarded in the network, which forms "hot spots" that have high traffic load. These hot spots can effectively mislead the attacker and increase the difficulty for attackers to trace the true sink node.

Compared with LPR, DEFP provides better protection to sink location privacy due to mechanisms in eliminating temporal correlation in the network traffic. It also increases volumes of data transmitted, resulting in large energy consumption.

7.4.2.3 Dummy packets strategies

To further protect the location of the data source, fake data packets can be introduced to perturb the traffic patterns that can be observed by the attacker.

In addition to the random walk mechanism combined with fake packets mentioned above, the Cyclic Entrapment Method (CEM)[55] is another typical routing protocol that is based on fake packets.

CEM generates link loops in the network and misleads external attackers to these loops to protect the source location privacy. After the deployment of the network, every node generates a loop with a certain probability. When a node in a loop receives a data message from source nodes, it will send fake messages on the loop it is in. Because attackers cannot distinguish fake packets from true data packets, they may be misled to the loop and trace along the loop until arriving back to the true path. Therefore, it will take more time for the attackers to trace back to the source node.

Although CEM can obtain good safety period, the introduction of fake messages brings great energy waste. Moreover, the safety of CEM will be destroyed if the attacker has the ability to observe traffic in a large area or to record nodes it has visited.

7.4.2.4 Fake nodes strategies

By placing some fake nodes that imitate the behaviors of the protected nodes as proxy nodes, the attackers can be attracted to the proxy nodes which are far away from the true target nodes. With this mechanism, the location privacy of target nodes is protected. A typical such protocol is the decoy sink nodes protocol.

The decoy sink nodes protocol is proposed to protect sink node location using data fusion technology. In this protocol, multiple faked sink nodes are deployed in the network. Collected data is firstly fused, and the results are passed to the decoy sinks. The decoy sink nodes perform further fusion on received data and pass the final result to the true sink node.

The decoy sinks shares the data flow to the real sink node. Due to the data fusion performed on decoy sink nodes, the volume of data sent from the decoy sinks to the real sink are not large. Hence the data packages received by the real sink are comparably equal with that of the decoy sinks, which eliminates the non-equivalences of communication patterns in the network.

In this protocol, the number of decoy sinks has great impact on the security of the network. When the number of decoy sinks is small, the attackers can still reveal the sink node's location privacy with high probability. Because decoy sinks are fixed, attacks launched to a decoy sink will incur a data loss of $1/N$ where N is the number of decoy sinks. When the number of decoy sinks is comparably large, the fusion function used needs to have a compress ratio of $1/N$ to ensure the equivalent of the real sink's traffic load and decoy sinks' traffic loads. This will cause information loss in some degree.

In addition to the above typical defense strategies, there are other strategies to protect the location privacy, including cross-layer solutions. Shao et al.[62] propose to use IEEE802.15.4 MAC layer beacon packages to protect the source location privacy. In this protocol, the source periodically broadcasts beacon packages (which are usually to declare some system parameters) containing real messages to be transmitted. The beacon packets will be first transmitted in the MAC layer for several hops and then transmitted to the sink node in the network layer using shortest path routing. Transmission in the MAC layer can well protect location privacy for the source node; but the cost is a higher propagation delay, because the interval between two successive beacon packet forwarding is relatively long.

7.4.3 Future research directions

Although many location privacy protection protocols have been proposed, there are still some open research issues to be solved. We list two potential research directions below.

(1) Most existing location privacy protection strategies depend on techniques such as random walk, decoy nodes, and fake packets. These techniques, however, usually cause high energy consumption and large transmission delay. On the other hand, existing researches either only consider protecting source locations or only consider protecting sink location. It is necessary and challenging to design and implement strategies that can simultaneously protect location privacies of the source and the sink nodes with low cost.

(2) In WSNs, protection of location privacies of mobile base stations is a challenging issue. It is obvious that a mobile base station can protect its location privacy well against external attackers; but it still needs to update its location information to the network. This may give more opportunities for internal attackers to trace it. It is an important open research issue to protect the mobile base station location privacy in order to ensure security.

7.5 Secure data aggregation

Data aggregation is a technique that can reduce the amount of transmitted data in WSNs by summarizing or combining raw readings from many sensor nodes. Sensor nodes are usually densely deployed in target districts; so the data collected by nearby nodes are usually redundant, both spatially and temporally. Data aggregation protocols leverage the redundancy of data to combine or compress readings from different nodes. This reduces the amount of data transmission meanwhile retaining required information. Aggregating data can reduce data transmission in the network, improving energy efficiency and bandwidth utility.

Data aggregation also negatively affects some performance metrics[63–65]: It may increase data transmission delay, degrade the accuracy of collected data, and increase the vulnerability of the whole network. Because WSNs are usually deployed in hostile environments, they require a high level of security. This goal of securing WSNs conflicts with the goal of data aggregation.

The former requires encrypting or authenticating data packets transmitted between neighboring nodes to provide security, while the latter requires plain data to perform aggregation efficiently. Generally, data aggregation cannot be performed on encrypted data. In order to perform data aggregation, intermediate aggregators need to decrypt received data first, then perform aggregation and encrypt the result before relaying the result to other nodes. This decryption-aggregation-encryption procedure makes the network more vulnerable to attacks. If an aggregator operating on readings from many sensor nodes were compromised, the adversary could forge or alter the aggregation result in arbitrary ways which could damage the final aggregating result. Furthermore, this procedure also exposures confidentiality of data and incurs additional computational overhead which degrade the efficiency of data aggregation protocols. Thus it is a critical research issue to provide secure data aggregation protocols in energy-efficient manners in WSNs.

7.5.1 Security requirements in data aggregation protocols

The logical topologies used in different data aggregation protocols are diverse. According to the number of aggregator layers used, typical data aggregation protocols can be classified into two categories: single layer aggregators and multiple layer aggregators. For different types of data aggregation protocols, the methods to achieve required security level are diverse. In general, a secure data aggregation protocol needs to provide some or all of the security requirements listed below[64].

1. *Data confidentiality*

Data confidentiality means that nodes' sent data is not disclosed to unauthorized users. Providing data confidentiality is the most important issue in mission critical applications. Due to the natural broadcasting property of wireless channels, in WSNs packets sent by a node can be heard by all its neighbors. In order to provide data confidentiality between two nodes, the transmitted packets need to be encrypted with keys only known by the two parties involving the communication. In most existing data aggregation protocols, aggregators cannot aggregate encrypted data directly; they need to decrypt received data first before performing aggregation. They also need to encrypt the aggregated results before sending them to the base station. This three-fold encryption-aggregation-decryption procedure increases not only transmission delay and computational overhead, but also the probability of aggregation protocols being attacked.

2. *Data integrity and freshness*

Data integrity means that the data used in aggregation is not altered or forged by adversaries. Data confidentiality guarantees that only authorized parties can obtain the data, but it cannot prevent the data from being corrupted. A compromised aggregator can alter the aggregation result or forge a false result to ruin the data integrity. The general methods to provide data integrity is to use message authentication code (MAC) or cyclic code. Furthermore, it is not enough to provide mere data integrity in WSNs. Compromised sensor nodes may listen to transmitted messages and launch replay attacks, which could disrupt the final aggregation result. Thus it is important to provide data freshness in data aggregation protocols against replay attacks.

3. *Source authentication*

With source authentication, a node can ensure that the node is communicating with is not a masqueraded node. Source authentication is mainly used to cope with Sybil attacks, in which a compromised node sends data to its aggregator under several fake identifies to disrupt the aggregation result. For the case when only two nodes communicate with each other, symmetric key encryption can be used to provide source authentication. For the case in which more than two nodes are involved in the communication (e.g., broadcasting), protocols such as μTESLA may be needed.

4. *Network/Service availability*

Network/service availability means that the network or the services provided by the network are still available under Denial-of-Services (DoS) attacks. An adversary may launch DoS attacks to some targeting nodes to prevent them from providing declared service. For data aggregation protocols, their function could be disrupted if aggregator nodes are targeted by DoS attacks. Thus it is important to guarantee availability of these aggregators in data aggregation protocols.

7.5.2 Secure data aggregation protocols

7.5.2.1 Overview

1. *Logical topologies in data aggregation protocols*

We divide the logical topologies used in typical data aggregation protocols into two categories: those that use single layer aggregators and those that use multiple layer aggregators, as shown in Fig. 7.4. In protocols that use single layer aggregators, sensor nodes send their raw readings to their aggregators, which then perform data aggregation and send the aggregation results to the base station. The routes from aggregator nodes to the base station may be single-hopped or multiple-hopped, but the aggregation results submitted by aggregators will not be aggregated again by other aggregators or sensor nodes en route. In protocols that use multiple layer aggregators, an aggregator may perform further aggregation on results from other aggregators. The main difference between the two types of data aggregation protocols are as follows. In protocols that use multiple layer aggregators, the aggregators in higher layers (those close to the base station in the topology) represent data from a great deal of sensor nodes in the network. This may essentially disrupt the final aggregation result if compromised by adversaries. On the other hand, sensor nodes or aggregators in lower layers (far from the base station in the topology) represent data from only a small part of the network, and the final aggregation results will not be affected much even if they are compromised.

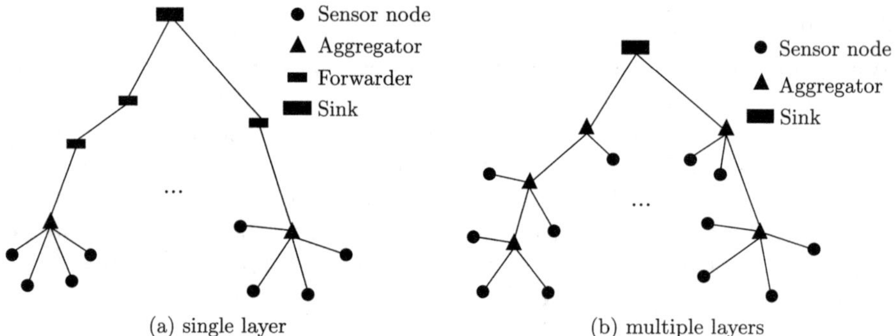

Fig. 7.4 Logical topologies used in data aggregation protocols.

Traditionally, logical topologies used in data aggregation protocols are classified into tree-based and cluster-based. This is different from our classification here. We argue that, compared with traditional classification method, our method is easy to understand. It demonstrates why it is difficult to provide end-to-end data confidentiality and why it is proposed to provide different level of security in some secure data aggregation protocols. We point out here that tree-based data aggregation protocols usually use multiple layer aggregators, while cluster-based data aggregation protocols can use either single layer aggregators or multiple layer aggregators.

2. *General techniques to provide data confidentiality*

Hop-by-hop data confidentiality can be achieved by encrypting messages transmitted between two communicating nodes with shared keys. The techniques to provide end-to-end (sensor nodes or aggregators to the base station) data confidentiality are diverse. In protocols that use single layer aggregators, the aggregators can encrypt their aggregation results using encryption keys given by the base station while the base station can decrypt the received messages and get the aggregation results. This is because in logical topologies using single layer aggregators, intermediate aggregator nodes do not need to know the content of packets form other aggregators. They do not need to decrypt packets from other aggregator, only forward them to the base station. In protocols that use multiple layers of aggregators, intermediate aggregator nodes need to perform further aggregation on data from other aggregators, which requires decrypting data from other aggregators first. The decryption-aggregation-encryption procedure incurs additional computational overhead as well as ruins end-to-end data confidentiality. In order to provide end-to-end data confidentiality in protocols with multiple layers of aggregators, privacy homomorphic cryptography has been used. With homomorphic cryptography systems, an intermediate aggregator node can directly aggregate on encrypted data without decrypting the data first. Thus sensor nodes or aggregators can encrypt their data with keys shared with the base station. Only the base station can decrypt the received message and intermediate aggregators cannot know the content of packets because they don't have decryption keys. With this method, end-to-end data confidentiality can be guaranteed in protocols using multiple layers of aggregators. The detailed description of privacy homomorphism is given in Section 7.5.2.3.

3. *General techniques to detect data alteration/forgery*

A compromised aggregator may forge data from sensor nodes that don't exist or alter data from authenticated sensor nodes to disrupt the final aggregation result. When receiving aggregation results from aggregators, the base station should have some mechanisms to detect these events and guarantee that the final aggregation results reflect the true readings sent by sensor nodes. A common method that can be used to detect forged or altered data is to commit to the data involved in the aggregation using the Merkle hash tree.

Fig. 7.5 shows a Merkle hash tree built on readings from eight sensor nodes. A Merkle tree is a binary tree in which the leaf nodes represent the hash value of raw readings of sensor nodes. Every intermediate node represents the hash value of the concatenation of its children. The root of the tree is called the commitment of the values represented by the leaves. The hash function used in the construction of the Merkle tree is collision resistant. When a Merkle tree is constructed, changes of values of any nodes in the tree will make the commitment change.

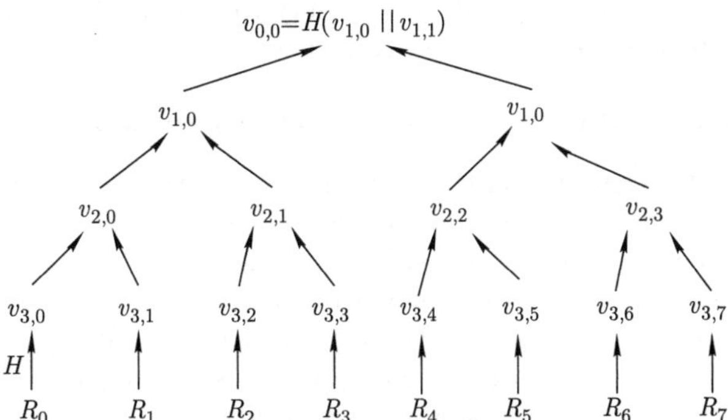

Fig. 7.5 Merkle hash tree.

When an aggregator sends its aggregation result to the base station, the commitment of the readings involved in the aggregation is also sent to the base station. The base station can check whether the aggregation result is derived from readings of corresponding sensor nodes, i.e., whether the aggregator used forged or altered data in the aggregation process. The procedure is as follows.

When the base station tries to check whether the reading from a sensor node is used in the aggregation, it requires the readings from that sensor node and the values on the verification path of the leaf node corresponding to the readings in the Merkle tree. The base station computes the commitment using these values and compares the computed value with the commitment received from the aggregator. If the two values match, the reading of the verifying sensor node is used in the aggregation; otherwise the reading of the verifying sensor node is forged or altered by the aggregator in the aggregation procedure. For example, the base station wants to verify whether the reading of sensor node n_0, say R_0, is correctly used in the aggregation. It first obtains R_0 from n_0 and computes corresponding hash value $H(R_0)$. The base station can guarantee that the reading it obtained is sent by n_0 with MAC or source authentication. The base station then requires the aggregator send the values on the verification path of R_0, i.e., the values of $V_{3,1}$, $V_{2,1}$, and $V_{1,1}$ to it. The base station can compute the commitment with $H(H(H(H(R_0)|V_{3,1})|V_{2,1})|V_{1,1})$ and compare this value with the value sent by the aggregator. If no match, it can be concluded that R_0 is not correctly used by the aggregator in the aggregation procedure.

The following two subsections introduce main secure data aggregation protocols developed in recent years. In data aggregation protocols that use single layer of aggregators, end-to-end data confidentiality can be provided using traditional cryptography systems. Thus in this type of protocols aggregation is usually operated on plain data. In data aggregation protocols that use multiple layers of aggregators, privacy homomorphic cryptography systems are needed to provide end-to-end data confidentiality. Thus we classify secure data aggregation protocols into two categories, protocols operating on plain data and protocols operating on encrypted data. This classification is consistent with those introduced in reference[63,64].

7.5.2.2 Secure data aggregation operating on plain data

The Secure Data Aggregation (SDA)[66] protocol proposed by Hu and Evans is the first secure data aggregation protocol. It is a tree-based protocol which uses multiple layers of aggregators. It assumes all the nodes in the network form a data collection tree in which the leaf nodes are sensing nodes and other nodes are aggregators. The key idea of SDA is to delay aggregation to the second hop in order to prevent a compromised aggregator from dropping, altering or forging immediate aggregation results. In the protocol, every leaf node generates a MAC using its shared key with the base station and sends its identification, its reading and the MAC to its parent node. Instead of performing aggregation immediately, the parent node forwards the received message to its own parent node which will perform aggregation on the data received from its grandchildren nodes. The parent node should also buffer the data received from its children for later verification. When the base station receives the aggregation results, it broadcasts authentication keys so that every aggregator can verify the message it receives from its children. If a leaf node is compromised and its reading is modified, the final aggregation result will be only slightly affected. If an aggregator is compromised and sends false aggregation result to its parent, the parent will be able to detect this event because the parent node has all readings of grandchildren. In SDA the base station uses μTESLA to update its shared keys with nodes. Using different keys in different round can counteract reply attacks thus provides data freshness. SDA provides data integrity when there is only one node being compromised, but it doesn't provide data confidentiality. It cannot cope with the cases when a node and its parent node are both compromised.

Przydatek et al. proposed the SIA protocol[67] which mainly targets on stealthy attacks. SIA uses single layer of aggregators. The authors assume there are three types of nodes in the network: a home server, an aggregator, and sensor nodes. They proposed an aggregate-commit-proof framework to verify whether the aggregation result submitted by the aggregator is a good approximation of the true value. There are three steps in this proposed framework. First, all sensor nodes send their readings to the aggregator. The aggregator then aggregates the readings and commits to these values using Merkle hash tree described in Section 7.5.2.1. When the base station receives the aggregation result committed by the aggregator, it verifies the confidentiality of result by randomly sampling raw readings in an interactive proof. SIA provides the following guarantees: if the aggregation result submitted by the aggregator is approximate to the ground-truth value, this result has high probability to be accepted by the base station; if the result is rejected by the base station, with high probability it is far away from ground-truth value.

The procedure of random sampling in the interactive proof is as follows. It is assumed that every node in the network has pair-wise keys with the aggregator and the home server. When a sensor node reports its data to the aggregator, the aggregator can authenticate the sensor node with its corresponding MAC. The aggregator constructs a Merkle hash tree using the data sent by sensor nodes. When the aggregator reports the aggregation result to the home server, it also sends the commitment of the data involved in aggregation (the value of the root of the Merkle hash tree) to the home server. The home server randomly selects some leaf nodes to check if their readings are correctly used in the aggregation. If the aggregator modifies or forges data in aggregation, the probability that this misbehavior could pass the verification is small. Furthermore, the home server can adaptively adjust the sampling rate to reduce this probability. Sybil attacks can also be detected by first sorting leaf nodes when constructing the Merkle hash tree and then sampling on two consecutive leaf nodes. At last the home server computes the probability of the reported aggregation result to see whether it is within a threshold of the ground-truth value and decides to accept or reject the value.

SIA provides data confidentiality, integrity, and source authentication. However, it cannot detect if a compromised node sends forged readings to the aggregator. This protocol can only provide probabilistic guarantee on the truth of the accepted aggregation result.

SecureDAV, proposed by Mahimkar and Rappaport[68], is a cluster-based secure data aggregation which uses a single layer of aggregators. There are two steps in the protocol cluster key establishment and aggregation result verification. For cluster key establishment, the authors proposed to use Elliptic Curve Cryptosystems (ECC) to generate a cluster key for each cluster. Compared with RSA, ECC can achieve the same security level with shorter keys. Furthermore, ECC incurs less computational overhead than RSA, thus is more efficient in energy and storage and is more suitable for WSNs. SecureDAV uses the (t, n)-threshold secret sharing mechanism to generate cluster keys. With this mechanism, every node in a cluster only knows a part of the cluster key, which guarantees that the cluster key cannot be revealed by the adversary when there are less than t nodes being compromised. Thus this mechanism provides data confidentiality in each cluster.

In the data aggregation and verification phase, the cluster head in each cluster first collects data from sensor nodes and performs aggregation. It then broadcasts the aggregation result to all the members in the cluster. Upon receiving the aggregation result, every sensor node generates a partial signature on the aggregation result and sends the signature to the cluster head. The cluster head combines all partial signatures into a whole signature and sends it to the base station along with the aggregation result. The base station authenticates the signature with its private key. If the cluster head is compromised, it cannot forge the signature because it doesn't know the cluster key. Furthermore, when the number of compromised nodes is less than t in a cluster, the adversary cannot reveal the cluster key. This protocol can counteract collusion attack in some degree. SecureDAV also employs a Merkle hash tree to detect modified or forged data used in aggregation. SecureDAV uses asymmetric cryptography systems to encrypt messages thus has high requirements on hardware.

Du et al. proposed the WDA[69] protocol which uses witness nodes to verify the correctness of the aggregation result submitted by an aggregator. For every aggregator, there are witness nodes which verify if the aggregator submits the correct aggregation result. Witness nodes collect the same data as corresponding aggregator and also perform aggregation on the data. However, they don't send the aggregation result to the base station. They compute MAC of the aggregation result and send MACs to the aggregator. When the aggregator reports its aggregation result to the base station, it must also send MACs from its witness nodes as the evidence of the correctness of the aggregation result. The base station uses a voting mechanism to check whether the aggregation result is correct. Assuming there are m witness nodes, the base station uses $n + 1$ out of m voting to check the correctness of the aggregation result. If more than n MACs from witness nodes are right, the base station accepts the aggregation result. If this verification fails, the base station will poll witness nodes to get the correct aggregation result. The authors analyzed the average number of rounds needed for the base station to get a correct result from the aggregator or witness nodes or to assert the unavailability of correct results due to lack of honest witness nodes. The WDA protocol provides data integrity but cannot guarantee data confidentiality. Furthermore, this protocol cannot counteract collusion attacks, in which the aggregator and corresponding witness nodes collude to cheat the base station. In WDA, an aggregator needs to forward MACs from its witness nodes to the base station, which incurs high communication overhead.

The SDAP[70] protocol proposed by Yang et al. is a hop-by-hop secure data aggregation protocol. It is a tree-based protocol and uses multiple layers of aggregators logically. The authors argue that in a tree-based topology, aggregators that are near the base station represent data from a large part of nodes in the whole network. If they are compromised, the final aggregation result at the base station will be greatly affected. Thus a data aggregation protocol should provide high security to these aggregators. On the other hand, all the nodes in the network are the same in the sense that they only have simple and resource-restricted hardware; thus there is no reason to require nodes to undertake more responsibility and to be more trustworthy. Based on this, the authors proposed a method to divide sensor nodes in the network into equal-sized logical groups. After the partition, aggregators that are near the base station only aggregate data from a small part of nodes in the network, reducing the damage to the final aggregation result if they are compromised. In order to still benefit the high energy efficiency from hop-by-hop aggregation mechanism, SDAP performs hop-by-hop data aggregation in each logical group.

In SDAP, after the base station receives all aggregation results reported by cluster headers of logical groups, it identifies those suspected results using a bivariate multiple-outlier detection algorithm. The basic idea is to use the Grubbs's test to detect outlier data (the authors extended the Grubbs's test such that it can detect outliers in data with two variables). The suspected logical groups need to be involved in an attestation procedure to prove the correctness of the reported aggregation result. The attestation procedure uses a method similar to Merkle hash tree to verify the correctness of a result. A randomly selected subset of nodes send their readings back to the base station. The base station computes the results and compares with the data reported by aggregators. The aggregation results reported from suspected logical groups that failed in the attestation procedure will be discarded. SDAP provides data confidentiality, integrity and source authentication.

The SRDA protocol[71] provides different levels of security to aggregators in different layers. It is a cluster-based data aggregation protocol. From the authors' point of view, in data aggregation protocols with multiple layers, messages transmitted between high layer aggregators present a combination of packets from many low level nodes. For example, for aggregation functions such as *max* and *count*, a partial result on higher level aggregators represent data from a large part of the network. Thus aggregators in higher layers should be guaranteed higher security. On the other hand, nodes in lower layers can be guaranteed with relatively lower security because the result will not be damaged much if they are compromised. In SRDA this is achieved by using RC6, a cryptosystem that can provide different security levels by adjusting the execution rounds. Furthermore, in order to reduce communication cost, in SRDA a node reports only the difference between its reading and reference data instead of reporting the raw data. SRDA uses a pre-distribution mechanism of keys, which improves efficiency by using location information of sensor nodes.

7.5.2.3 Secure data aggregation operating on encrypted data

The protocols discussed in the previous section all need to operate on plain data. In order to provide end-to-end data confidentiality, encryption-aggregation-decryption operations need to be performed on intermediate aggregators in the protocols. In this section, we introduce some secure data aggregation protocols that can operate on encrypted data directly. This is usually achieved by using privacy homomorphic cryptography.

Privacy homomorphism is an encryption transformation that allows direct aggregation on encrypted data. Let D and E be the decryption process and encryption process, respectively. Assume Kpr and Kpu are the base station's encryption key and decryption key, respectively. A privacy homomorphism is called additively homomorphic if

$$a + b = D_{Kpr}(E_{Kpu}(a) + E_{Kpu}(b)), \quad \text{where} \quad a, b \in Q,$$

and it is called multiplicatively homomorphic if

$$a \times b = D_{Kpr}(E_{Kpu}(a) \times E_{Kpu}(b)), \quad \text{where} \quad a, b \in Q.$$

The widely used RSA cryptosystem is a privacy homomorphism that is multiplicatively homomorphic. Generally speaking, the more operations a privacy homomorphism supports, the more computation sensitive it will be.

The canceled data aggregation (CDA) protocol[72] proposed by Westhoff et al. is a protocol that provides end-to-end data confidentiality with privacy homomorphism. It uses a single layer of aggregators. CDA employs an encryption function called Domingo-Ferrer approach, which is both additively and multiplicatively homomorphic. In CDA, before sending its reading to the aggregator, each sensor node encrypts the data using the key it shares with the base station. The aggregators perform aggregation directly on the encrypted data and send the intermediate aggregation results to the base station. The base station decrypts and computes the aggregation result after it receives all intermediate aggregation results. Because the aggregators do not have the knowledge of the decryption keys of sensor nodes, end-to-end data confidentiality is guaranteed. Compared with hop-by-hop data aggregation mechanisms, this mechanism is more flexible.

The drawback of CDA is that the Domingo-Ferrer function is a symmetric cryptography system which is vulnerable to plaintext attacks. The authors argue that, compared with the heavy overhead to successfully attach this cryptography system, the obtained information will be less valuable. The Domingo-Ferrer function is also very computation sensitive. In order to enhance security, in CDA a sensor node first divides its data into $d(2 \leqslant d \leqslant 4)$ small divisors before encrypting the data. This incurs both additional computational and communication cost. The Domingo-Ferrer function only supports additive and multiplicative homomorphism, thus cannot support some frequently used aggregation functions in data aggregation protocols that use multiple layers of aggregators such as median, min or max.

S. Ozdemir et al. proposed the CDAP protocol which takes advantage of asymmetric cryptography based privacy homomorphism system to enhance the security of data aggregation protocols operating on encrypted data. CDAP uses multiple layers of aggregators. It employs asymmetric cryptography based privacy homomorphism to provide end-to-end data confidentiality. However, asymmetric cryptography based privacy homomorphism incurs very high computational overhead that is unaffordable to simple sensor nodes with restricted resources. Thus the authors proposed to use special nodes that have rich resources (such as Intel's Stargate and iMote) as aggregator nodes to perform aggregation on encrypted data. In CDAP, aggregators share session keys with the base station. The sensor nodes transmit their encrypted data to aggregators. An aggregator first decrypts the data and performs data aggregation. It then encrypts the result using privacy homomorphism and sends the encrypted result to the base station. Intermediate aggregators can aggregate the result further upon receiving the encrypted data. At last, the base station decrypts the final result with its private keys. The drawback of this protocol is that it needs special hardware as aggregators, which incurs high cost to construct the network and reduces the flexibility of the topology.

7.5.3 Future research directions

We list some potential research directions below.

(1) A compromised aggregator node can inject forged data into the network to affect the final aggregation results. Because the main task of aggregators is to aggregate data from other nodes, it is difficult to detect whether the aggregator has injected false data into the network. Designing algorithms to detect this misbehavior effectively is a potential research issue.

(2) Privacy homomorphism provides mechanisms to support direct aggregating operation on encrypted data and to guarantee end-to-end data confidentiality. However, symmetric key based privacy homomorphism cannot provide enough security and asymmetric key based privacy homomorphism usually incurs high computational overhead which cannot be afforded by sensor nodes. It is necessary to find privacy homomorphism more suitable to

WSNs, e.g., those can provide enough security with affordable computational cost.

(3) Currently privacy homomorphism only supports limited aggregating operations, mainly those depending only on addition and multiplication (e.g., sum, average). However, the aggregating operations needed in WSNs applications are diverse. Thus we need to design privacy homomorphism that can support more types of operations.

7.6 Conclusion

Due to the critical role that security plays in WSN applications, construction of secure WSNs and how to enhance the security of protocols designed for WSNs is a hot research topic in recent years. A lot of works have been devoted to establishing secure WSNs and designing secure protocols for WSNs. In this chapter, we surveyed state-of-the-art solutions to some of these security issues in WSNs, mainly focusing on how to effectively and efficiently distribute and manage keys to provide link-wised communication security. We also discussed how to design secure higher layer protocols, including routing layer protocols and application layer protocols such as location privacy protection and secure data aggregation. Although great progress has been achieved in current WSNs security field, there are still many unsolved problems in each topic we discussed in this chapter. Further research is required in the potential directions listed for each topic in this chapter.

Acknowledgments

The authors would like to thank the anonymous reviewers and the editors for their invaluable suggestions that improved the quality of this manuscript. This work has been supported in part by the National Natural Science Foundation of China under Grant No. 60873265, 61103203, and 61173169, the Program for New Century Excellent Talents in University of Ministry under Grant No. NCET-10-0798, and the Postdoc program of Central South University.

References

[1] Xiao Y, Rayi V K, Sun B, Du X J, Hu F, Galloway M (2007) A survey of key management schemes in WSNs. Computer Communications 30(11 – 12): 2314 – 2341. doi:10.1016/j.comcom.

[2] Perrig A, Szewczyk R, Tygar J D, Wen V, Culler D E (2002) SPINS: Security protocols for sensor networks. Wireless Networks 8(5): 521 – 534.

[3] Eschenauerl L, Gligor V D (2002) A key-management scheme for distributed sensor networks. In Proceedings of the 9th ACM Conference on Computer and Communications Security (CCS), pp. 41 – 47.

[4] Chan H, Perrig A, Song D (2003) Random key pre-distribution schemes for sensor networks. In Proceedings of the 2003 IEEE Symposium on Security and Privacy, pp. 197 – 213.

[5] Zhu S, Setia S, Jajodia S (2003) LEAP: Efficient security mechanisms for large-scale distributed sensor networks. In Proceedings of the 10th ACM Conf. on Computer and Communications Security, pp. 62 – 72.

[6] Eltoweissy M, Mohamim M, Mukkamala R (2006) Dynamic Key Management in Sensor Networks. IEEE Communications Magazine, 44(4): 122 – 130.

[7] Blom R (1984) An Optimal Class of Symmetric Key Generation Systems. In Advances in Cryptology (EUROCRYPT), LNCS 209: 35 – 338.

[8] Blundo C, Santis A D, Herzberg A, Kutten S, Vaccaro U, Yung M (1992) Perfectly Secure Key Distribution for Dynamic Conferences. In Advances in Cryptology (CRYPTO), pp. 471 – 486.

[9] Du W, Deng J, Han Y S, Varshney P K, Katz J, Khalili A (2003) A Pair-wise Key Pre-distribution Scheme for Wireless Sensor Networks. In Proceedings of the 10th ACM Conf on Computer and Communications Security (CCS), pp. 42 – 51.

[10] Liu D, Ning P, Li R F (2003) Establishing Pairwise Keys in Distributed Sensor Networks. In Proceeding of the 10th ACM Conference on Computer and Communications Security (CCS), pp. 52 – 61.

[11] Liu D, Ning P (2003) Location-based pairwise key establishments for static sensor networks. In Proceedings of the 1st ACM Workshop on Security of Ad Hoc and Sensor Networks, pp. 72 – 82.

[12] Liu Z H, Ma J F, Huang Q P (2006) Domain-based Key Management for WSNs. Chinese Journal of Computers, 29(9): 1608 – 1616.

[13] Du W, Deng J, Han Y S, Chen S G, Varshney P K (2004) A key management scheme for WSNs using deployment knowledge. In Proceedings of IEEE INFOCOM, pp. 586 – 597.

[14] Chan H, Perrig A (2005) PIKE: Peer Intermediaries for Key Establishment in Sensor Networks. In Proceedings of IEEE INFOCOM, pp. 524 – 535.

[15] Traynor P, Choi H, Cao G H, Zhu S C, Porta T L (2006) Establishing Pair-wise Keys in Heterogeneous Sensor Networks. In Proceedings of IEEE INFOCOM, pp. 1 – 12.

[16] Younis M F, Ghumman K, Eltoweissy M (2006) Location-Aware Combinatorial Key Management Scheme for Clustered Sensor Networks. IEEE Transactions on Parallel and Distributed Systems, 17(8): 865 – 882.

[17] Kong F R, Li C W (2010) Dynamic key management scheme for wireless sensor network. Journal of Software, 21(7): 1679 – 1691.

[18] Eltoweissy M, Heydari M H, Morales L, Sadborough I H (2004) Combinatorial Optimization of Key Management in Group Communications. Journal of Network and System Management, 12(l): 33 – 50.

[19] Liu Z H, Ma J F, Huang Q P, Moon S J (2009) Asymmetric Key Predistribution Schemes for Sensor Networks. IEEE Transactions on Wireless Communications, 8(3): 1366 – 1372.

[20] Weiping W, Jinhong X, Jianxin W (2009)Detection and location of malicious nodes based on source coding and multi-path transmission in WSN. In Proceedings of the 11th IEEE International Conference on High Performance Computing and Communications (HPCC), pp. 458 – 463.

[21] Ana Paula S, Marcelo H.T. M, Bruno P.S. R, Antonio A.F. L, Linnyer B. R, Hao Chi W (2005)Decentralized intrusion detection in WSNs. In Proceedings of the 1st ACM International Workshop on Quality of Service & Security in Wireless and Mobile Networks, pp. 16 – 23.

[22] Zhiyuan G, Qiyuan H (2006) Study on Security of Routing Protocol in WSNs. Radio Engineering of China, 36(1): 17 – 20.

[23] Boping Q, Xianwei Z, Jun Y, Cunyi S (2006) Research on Secure Routing Techniques in WSNs. Chinese Journal of Sensors and Actuators, 19(1): 16 – 19.

[24] Karlof C, Wagner D (2003) Secure routing in WSNs: attacks and countermeasures. Ad Hoc Networks, 1(3): 293 – 315.

[25] Qinghua Z, Pan W, Douglas S. R, Peng N (2005) Defending against Sybil Attacks in Sensor Networks.In Proceedings of the 2nd International Workshop on Security in Distributed Computing Systems(ICDCS Workshop), pp. 185 – 191.

[26] James N, Elaine S, Dawn S, Adrian P (2004) The Sybil Attack in Sensor Networks: Analysis and Defenses. In Proceedings of the 3rd International Symposium on Information Processing in Sensor Networks (IPSN), pp. 1 – 8.

[27] Weimin L, Zongkai Y, Shizhong W, Yunmeng T (2005) Research on the Security in WSNs. Computer Science, 32(5): 54 – 58.

[28] Ngai E.C. H, Jiangchuan Liu, Lyu M. R (2006) On the Intruder Detection for Sinkhole Attack in WSNs. In Proceedings of IEEE International Conference on Communication (ICC), pp. 3383 – 3389.

[29] Yanchao Z, Wei L, Wenjing L, Yuguang F (2005) Location-based compromise-tolerant security mechanisms for WSNs. IEEE Journal on Selected Areas in Communications, 24(2): 247 – 260.

[30] Benjamin J. C, H. Chris T (2004) Sinkhole intrusion indicators in DSR MANETs. In Proceedings of the first international conference on Broadband Networks (BroadNets), pp. 681 – 688.

[31] Yong T, Mingtian Z, Xin Z (2006) Overview of Routing Protocols in WSNs. Journal of Software, 17(3): 410 – 421.

[32] Liang D, Xiaohui C, Wentao W (2009) Research on routing protocol for WSNs. Sensor World, 15(9): 26 – 29.

[33] Junlei B, Xinhui R, Zhengwei G (2008) Research on Rouitng Protocol Classification for WSNs. Computer Technology and Development, 18(5): 131 – 134.

[34] Chalermek I, Ramesh G, Deborah E (2000) Directed diffusion: a scalable and robust communication paradigm for sensor networks. In Proceedings of the 6th Annual ACM/ IEEE International Conference on Mobile Computing and Networking (MobiCom), pp. 56 – 67.

[35] David B, Deborah E (2002) Rumor routing algorithm for sensor networks. In Proceedings of the 1st ACM international workshop on WSNs and applications (WSNA), pp. 22 – 31.

[36] Wendi R H, Anantha C, Hari B (2000) Energy-efficient communication protocol for wireless microsensor networks. In Proceedings of 33rd Annual Hawaii International Conference on System Sciences, pp. 3005 – 3014.

[37] Arati M, Dharma P A (2001) TEEN: A Routing Protocol for Enhanced Efficiency in WSNs. In Proceedings of 15th International Parallel and Distributed Processing Symposium (IPDPS) 1: 2009 – 2015.

[38] Lindsey S, Raghavendra C S (2002) PEGASIS: Power-efficient gathering in sensor information systems. In Proceedings of IEEE Aerospace Conference: 1125 – 1130. doi: 10.1109/AERO.2002.1035242.

[39] Yan Y, Remesh G, Deborah E (2001) Geographical and energy aware routing: a recursive data dissemination protocol for WSNs. UCLA Computer Science Department Technical Report: UCLA/CSD-TR-01-0023. doi: 10.1.1.21.8533.

[40] Brad K, H.T. K (2000) GPSR: greedy perimeter stateless routing for wireless networks. In Proceedings of the 6th annual international conference on Mobile Computing and Networking (Mobicom), pp. 243 – 254.

[41] Benjie C, Kyle J, Hari B, Robert M (2002) Span: an energy-efficient coordination algorithm for topology maintenance in Ad Hoc wireless networks. Wireless Networks Journal, 8: 481 – 494.

[42] Ya X, John H, Deborah E (2001) Geography-informed energy conservation for ad hoc routing, In Proceedings of the 7th Annual International Conference on Mobile Computing and Networking (Mobicom), pp. 70 – 84.

[43] Yang Z, Zhi-Hua F, Xiao-Xin H, Yu-Xin W (2005) Anonymous Secure Multipath Routing in Mobile Ad Hoc Networks. Acta Electronica Sinica, 33(11): 2022 – 2030. doi: cnki:ISSN:0372-2112.0.2005-11-022.

[44] Yih-Chun H, Adrian P, David B. J (2003) Packet leashes: a defense against wormhole attacks in wireless ad hoc networks. In Proceedings of the 22nd Annual Joint Conference of the IEEE Computer and Communications (Infocom), pp. 1976 – 1986.

[45] Xia W, Johnny W (2007) An End-to-end Detection of Wormhole Attack in Wireless Ad Hoc Networks. In Proceedings of the 31st Annual International Computer Software and Applications Conference (COMPSAC)1: 1 – 8.

[46] Lingxian H, David E (2004) Using Directional Antennas to Prevent Wormhole Attacks. In Proceedings of the Network and Distributed System Security Symposium (NDSS), pp. 131 – 141.

[47] Weichao W, Bharat B (2004) Visualization of wormholes in sensor networks. In Proceedings of the 34th ACM workshop on Wireless security (WiSe), pp. 51 – 60.

[48] Issa K, Saurabh B, Ness .B. S (2005) LiteWorp: a lightweight countermeasure for the wormhole attack in multihop wireless networks, In Proceedings of the International Conference on Dependable Systems and Networks (DSN), pp. 612 – 621.

[49] Liang H, Fan H, Bing P, Jing C (2006) Defend against Wormhole Attack Based on Neighbor Trust Evaluation in MANET. Computer Science, 33(8): 130 – 138.

[50] Jing D, Richard H, Shivakant M (2006) INTRSN: Intrusion-tolerant routing in WSNs. Computer Communications, 29(2): 65 – 71.

[51] Zhen C, Jianbin H, Zhong C, Maoxing X, Xia Z (2006) Feedback: Towards Dynamic Behavior and Secure Routing for WSNs. In Proceedings of the 20th International Conference on Advanced Information Networking and Applications (AINA), pp. 1 – 5.

[52] Xiao-Yun W, Li-Zhen Y, Ke-Fei C (2005) SLEACH: Secure Low- Energy Adaptive Clustering Hierarchy Protocol for Wireless Sensor Networks. Wuhan University Journal of Natural Sciences, 10(1): 127 – 131.

[53] KAMAT P, Yanyong Z, TRAPPE W, Trappe W, Ozturk C (2005) Enhancing source-location privacy in sensor network routing. In Proceedings of the 25th International Conference on Distributed Computing Systems (ICDCS), pp. 599 – 608.

[54] Weiping W, Liang C, Jianxin W (2008) A source-location privacy protocol in WSN based on locational angle. In Proceedings of the 43th IEEE International Conference on Communication (ICC), pp. 1630 – 1634.

[55] Yi OY, Zhengyi L, Guanling C, James F, Fillia M (2006) Entrapping adversaries for source protection in sensor networks. In Proceedings of the 7th IEEE International Symposium on a World of Wireless, Mobile and Multimedia Networks (WOWMOM), pp. 23 – 32.

[56] Yong X, Schwiebert L, Weisong S (2006) Preserving source location privacy in monitoring-based wireless sensor networks. In Proceedings of the 20th Symposium on Parallel and Distributed Processing (IPDPS), pp. 1 – 8.

[57] Liang Z (2006) A self-adjusting directed random walk approach for enhancing source-location privacy in sensor network routing. In Proceedings of the 2nd International Conference on Communications and Mobile Computing (IWCMC), pp. 33 – 38.

[58] Conner W, Abdelzaher T, Nahrstedt K (2006) Using data aggregation to prevent traffic analysis in wireless sensor networks. In Proceedings of the 2nd IEEE International Conference on Distributed Computing in Sensor Systems (DCOSS), pp. 202 – 217.

[59] Ying J, Shigang C, Zhan Z et al (2007) Protecting receiver-location privacy in wireless sensor networks. In Proceedings of the 26th IEEE International Conference on Computer Communications (INFOCOM), pp. 1955 – 1963.

[60] Deng J, Han R, Mishra S (2006) Decorrelating wireless sensor network traffic to inhibit traffic analysis attacks. Pervasive and Mobile Computing, 2(2): 159 – 186.

[61] Ozturk C, Yanyong Z, Frappe W et al (2004) Source-location privacy for networks of energy-constrained sensors. In Proceedings of the 2nd IEEE Workshop of Software Technologies for Future Embedded and Ubiquitous Systems (WSTFES), pp. 68 – 72.

[62] Min S, Wenhui H, Sencun Z et al (2009) Cross-layer enhanced source location privacy in sensor networks. In Proceedings of the 6th Annual IEEE Communications Society Conference on Sensor, Mesh and Ad Hoc Communications and Networks (SAHCN), pp. 1 – 9.

[63] Alzaid H, Foo H, Nieto J G (2008) Secure data aggregation in wireless sensor network: a survey. In Proceedings of the sixth Australasian conference on Information security (AISC), 85: 93 – 105.

[64] Ozdemir S, Yang X (2009) Secure data aggregation in WSNs: A comprehensive overview. Computer Networks, 53: 2022 – 2037.

[65] Yingpeng S, Hong S, Yasushi Ii, Yasuo T, Naixue X (2006) Secure data aggregation in WSNs: A survey. In Proceedings of the seventh International Conference on Parallel and Distributed Computing, Applications and Technologies (PDCAT), pp. 315 – 320.

[66] Lingxuan H, Evans D (2003) Secure aggregation for wireless networks. In Proceedings of the Symposium on Applications and the Internet Workshops (SAINT workshops), pp. 384 – 391.

[67] Bartosz P, Dawn S, Adrian P (2003) SIA: Secure information aggregation in sensor networks. In Proceedings of the first international conference on embedded networked sensor systems (Sensys), pp. 255 – 265.

[68] Mahimkar A, Rappaport TS (2004) SecureDAV: A secure data aggregation and verification protocol for sensor networks. In Proceedings of Global Telecommunications Conference (Globecom), pp. 2175 – 2179.

[69] Wenliang D, Jing D, Yunhgsiang S H, Pramod K V (2003) A witness-based approach for data fusion assurance in WSNs, In Proceedings of Global Telecommunications Conference (Globecom), 3: 1435 – 1439.

[70] Yi Y, Xinran W, Sencun Z, Guohong C (2008) SDAP: A secure hop-by-hop data aggregation protocol for sensor networks. ACM Transactions on Information and System Security (TISSEC), 11(4): 1 – 43.

[71] H. Ozgur S, Suat O, Hasan C (2004). SRDA: secure reference-based data aggregation protocol for WSNs. In Proceedings of the 60th Vehicular Technology Conference (VTC), 7: 4650 – 4654.

[72] Girao J, Westhoff D, Schneider M (2005) CDA: Concealed data aggregation for reverse multicast traffic in wireless sensor networks. In Proceedings of IEEE International Conference on Communications (ICC), 5: 3044 – 3049.

Chapter 8
Security in Wireless Sensor Networks

Ping Li[1,2], Limin Sun, Xiangyan Fu, and Lin Ning

Abstract

As wireless sensor networks edge closer towards wide-spread deployment, security issues become a central concern. However, the more challenging it becomes to fit the security of WSN into that constrained environment including very limited energy resources, low abilities to resist physical attacks, and lack of feedback mechanisms for abnormal cases off-line. Thus the research of security issues in WSN is very important. The intent of this chapter is to investigate the security related issues in wireless sensor networks. Firstly, the security architecture of sensor networks is proposed, trying to outline a general illustration on this area. Then, the following four aspects are investigated.

(1) The cryptographic mechanisms.
(2) Various keying mechanisms for the key management issue.
(3) A panoramic view and detailed analysis of the trust management.
(4) A set of effective strategies based on protecting location privacy.

8.1 Introduction

A wireless sensor network (WSN) consists of a set of compact and automated devices called sensor nodes. A typical sensor network has hundreds to millions of sensor nodes. Each sensor node is typically low-cost, limited in computation and information storage resource, highly power constrained, and communicates over a short-range wireless network interface. These features ensure a wide range of applications for sensor networks, including military provision, environment monitoring and exploring on man-unreachable circumstances[1].

1 Changsha University of Science & Technology, Changsha, Hunan, China, 410004.
2 Institute of Software, Chinese Academy of Science, Beijing, China, 100190.

It appears that the security issues of sensor networks have not been considered as sufficiently as it should be[2]. In many applications, environment monitoring and battlefield spying, for instance, the nodes are subject to attacks like passive eavesdropping, active intrusion, message flooding, fake information inserting, etc. In the above hostile attacks, passive eavesdropping helps adversaries intercept private information. Active intrusion makes it possible for adversaries to delete information, insert false information or impersonate nodes, which destroy the usability, integrality, security certificate and non-reputation of WSNs. In consequence, the security issues have gained much interest. Key management plays a very key role in deploying security strategies of sensor networks, including key pre-distribution, key discovery and key maintenance. However, the threats faced by WSNs are not only from external attackers, but also from internal nodes which are compromised as byzantine nodes, and some internal nodes may conduct selfishly for the sake of energy conservation. Comparing with external attacks, internal attacks are more difficult to defend because the key mechanisms are ineffective against internal malicious nodes, therefore internal attacks can make worse threats to the network. There needs to be urgent solutions for legitimate nodes to detect and further eliminate malicious nodes.

Trust management is essential for identifying malicious, selfish and compromised nodes which have been authenticated. It has been widely studied in many network environments such as peer-to-peer networks, grid and pervasive networks and so on. However, in reality, sensor nodes have limited resources and other special characteristics, which make trust management for WSNs more significant and challenging. Up to the present, research on the trust management mechanisms of WSNs have mainly focused on nodes' trust evaluation to enhance the security and robustness. The practical applications of this method include route, data integration and cluster head vote. Although some existing approaches play good roles in improving security of other networks, trust management in WSNs still remains to be a challenging field.

In addition, compared to traditional networks, WSNs are resource constrained and application specific, which determines that privacy problems are significantly distinguishable and unique, making it more difficult to effectively apply existing privacy protection mechanisms and algorithms to address related problems. Consequently, it brings emergent requirements and great challenges for designing privacy protection solutions within WSNs.

As the security issues cover many detailed topics, we propose the security architecture of sensor networks in this chapter and summarize current research achievements based on this architecture. We investigate the security issue in four aspects: cryptographic approaches, resilience on key management, trust management and location privacy mechanisms. The main reason for such a consideration is that there exists a fundamental contradiction between the origin of sensor networks and conventional security characteristics. Towards these issues, based on well-established mathematical models, we propose corresponding solutions, algorithms and protocols. The rest of this chapter is organized as follows. Section 8.2 gives an overview of security architecture for WSN. Section 8.3 presents discussions on cryptographic. Section 8.4 provides a detailed analysis of the resilience on key management issues; and the trust management of WSN is addressed in Section 8.5. Section 8.6 provides a set of nice strategies to protect location privacy. The chapter is concluded in Section 8.7.

8.2 Overview of security architecture for WSN

In this section, we describe the following aspects of sensor architecture for WSNs: various attacks on WSN nodes, security requirements, and hierarchical architecture for WSN security.

8.2.1 Malicious nodes attacks in WSNs

WSNs are particularly vulnerable to a variety of security threats, such as malicious nodes on the transmission paths dropping, fabricating, or tampering the forwarded messages, and denial of service, while prompting a range of fundamental research challenges. The typical attacks in wireless sensor network include wormhole attack, sinkhole attack and sybil attack and so on, in which malicious nodes always try to participate in a path or compromise the nodes on path, so as to drop, fabricate or tamper messages. There are many papers[3–9] that describe these security threats. We follow Anthony D. Wood's classification of attacks into different layers[10]. Each layer is susceptible to different attacks and has different options available for its defense. Some attacks crosscut multiple layers or exploit interactions between them.

8.2.1.1 Physical layer attacks

Since the use of technology of wireless communication in WSN, it is easily to incur jamming attack from attackers in physical layer. Moreover, physical access to the sensor node is possible because of the placement of sensor nodes in an unguarded environment. Therefore, an intruder may be able to tamper or damage with the sensor devices.

1. *Jamming*

As a well-known attack to wireless communications, jamming is one of many exploits used compromise the wireless environment. Jamming can be a huge problem for wireless networks, since radio frequency (RF) is essentially an open medium. Jamming can disrupt wireless transmission. And it can occur either unintentionally in the form of interference, noise or collision at the receiver side or in the context of an attack. Even sporadic jamming can be sufficient to cause disruption because the communication data carried by the network may be available for only a short time. This attack is very effective for single frequency networks. Adversaries can disrupt the network through launch radio waves near the frequency point, as long as they get the center frequency of communication frequency.

Conventional defense techniques against physical layer jamming rely on spread spectrum, which can be too energy-consuming to be widely deployed in resource constrained sensors. Mobile-phone networks generally use code spreading as a defense against jamming. In addition, when jamming is intermittent, nodes may be able to report the attack to the base station by sending a few high-power and high-priority messages. In order to maximize the probability of successfully delivering such messages, nodes should cooperate with each other, for example, switching to a prioritized transmission scheme that minimizes collisions. Nodes can also buffer high-priority messages indefinitely so as to relay them once a gap in the jamming occurs.

2. *Tampering*

An adversary can tamper with nodes physically, and interrogate and compromise them, which aggravates the threats of large-scale sensor networks. However, it is unpractical to control access to hundreds of nodes spread over several kilometers. Furthermore, an attacker may be able to destroy or replace the sensor and computational hardware, even extract sensitive materials such as encryption keys to get unlimited access to higher levels of communication. Therefore, such networks can fall prey to true brute-force destruction[11].

Focused on the dangers discussed above, one countermeasure called tamper-proofing is presented. Tamper-proofing is a method used to hinder, deter or detect unauthorized access to a device or circumvention of a security system. When possible, the node should respond to tampering in a fail-complete manner. For example, it could cryptographic or erase program memory. There also are many other traditional physical defenses such as camouflaging, hiding nodes and so on.

8.2.1.2 Link layer attacks and countermeasures

The link or Media Access Control (MAC) layer provides channel arbitration for neighbor-to-neighbor communication. Cooperative schemes that depend on carrier sense, which let nodes detect if other nodes are transmitting, are particularly vulnerable to all kinds of attacks. For example, collisions and unfairness at the link layer may be able to delay the packet transmission or cause the packet to be corrupted.

1. *Collision*

Suck attacks can be easily launched by a compromised (or hostile) sensor node. In a collision attack, an attacker node does not follow the medium access control protocol and cause collisions with neighbor node's transmissions by sending a short noise packet. This attack does not consume much energy of the attacker but can cause a lot of disruptions to the network operation. It is not trivial to identify the attacker due to the wireless broadcast nature. Adversaries may be able to disrupt an entire packet only need to induce a collision in one octet of a transmission.

These malicious collisions which create a kind of link-layer jamming can be identified by the network to use collision detection. However, this approach cannot completely effective defense this attack. Proper transmission still requires cooperation among nodes, which is expected to escape corruption of others' packet. A subverted node could repeatedly and intentionally deny access to the channel, expending much less energy than in full-time jamming.

2. *Unfairness*

This threat may not entirely prevent legitimate access to the channel and the use of small frames means that the channel is only captured for a small amount of time. However, the adversary could cheat by quickly responding when needing access while other nodes delay, for example, causing users of a real-time MAC protocol to miss their deadlines.

One method of defending against this threat is to use small frames so as to an individual node can only capture the channel for a short time. Nevertheless, this approach increases framing overhead if the network typically transmits long messages. Furthermore, when vying for access, an attacker can defeat this defense by cheating, such as by responding quickly while others delay randomly.

3. *Exhaustion*

As introduced in reference [10], this active attack may attempt retransmission repeatedly, even when attracted by an unusually late collision, such as a collision induced near the end of the frame. In nearby nodes, this threat could culminate when the battery resources was exhausted. A self-sacrificing node could develop the interactive nature of most MAC-layer protocols in an interrogation attack. For example, IEEE 802.11 which based MAC protocols uses request-to-send (RTS), clear-to-send (CTS), and Data/ACK messages to transmit data and reserve channel access. The node could elicit a CTS response from the targeted neighbor and repeatedly request channel access. Constant transmission would finally exhaust the energy resources of both nodes.

One countermeasure to prevent this attack is to makes the MAC admission control rate limiting, so that the network can ignore excessive requests without sending expensive radio transmissions. Nonetheless, this limit cannot drop below the expected maximum data rate the network supports. One design-time strategy for protection against battery-exhaustion attacks limits the extraneous responses the protocol requires. Designers usually code this capability into the system for general efficiency, but coding to handle possible attacks may require additional logic.

8.2.1.3 Network and routing layer attacks and countermeasures

Network layer attacks are a significant and credible threat to wireless sensor networks. This layer provides a critical service. Before reaching their destination, messages may pass through a lot of hops in a large-scale deployment. Unfortunately, as the aggregate network cost of relaying a packet increases, the probability of the dropping or misdirecting packet along the way in the network increases as well.

1. *Homing*

In the majority of sensor networks, some nodes will have special responsibilities, for example, they are elected the leader of a local group for coordination. More powerful nodes might serve as cryptographic key managers, monitoring access points or query, or network uplinks. Because these nodes provide critical services to the network, they often attract an adversary's interest. Location-based network protocols that rely on geographic forwarding[13] expose the network to homing attacks. Here, a passive adversary learns the presence and location of critical resources by observing traffic. Once found, its collaborators or mobile adversaries can attack these nodes by using other active means.

One effective approach to hiding significant nodes provides confidentiality for both message headers and their content. The network can encrypt the headers at each hop supposing that all neighbors share cryptographic keys. This would prevent a passive adversary from easily learning about the source or destination of overheard messages, if a node has not been subverted and remains in possession of valid decryption keys.

2. *Neglect and greed*

This threat is a simple form of attack arbitrarily neglects to route some messages to attacks the node-as-router vulnerability. In this kind of attack, the subverted or malicious node can still take part in lower-level protocols, and may even acknowledge reception of data to the sender, but it may refuse to forward packets or drop them on a random or arbitrary basis. Also, it can forward to packet to wrong receiver and gives undue and high priority to its own messages, so as to destroy the network communication rule. Furthermore, the dynamic source routing (DSR) protocol[12] is susceptible to this attack. Communications from a region may all use the same route to a destination as the network caches routes. If a node along that route is greedy, it may consistently degrade or block traffic from the region to a base station.

Multipath routing can be used to counter this type of attack. Messages routed over n paths whose nodes are completely disjoint are completely protected against neglect and greed attacks involving at most n compromised nodes and still offer some probabilistic protection when over n nodes are compromised. The use of multiple braided paths may provide probabilistic protection against selective forwarding and use only localized information. Allowing nodes to dynamically choose next hop from a set of possible candidates can further reduce the chances of an adversary gaining complete control of a data flow. Sending redundant messages is effective countermeasure. It is difficult to distinguish a greedy node from a failed node, however, so prevention is safer than relying on detection.

3. *Misdirection*

Misdirection is based upon changing, spoofing, or replaying the routing information. By forwarding the message along with the wrong path or by sending false routing updates can lead to this kind of attack. This attack targets the sender and diverts traffic away from its intended destination. Moreover, by misdirecting many traffic flows in one direction, this attack can target an arbitrary victim. In one variant of misdirection, Internet smurf attacks, the attacker forges the victim's address as the source of many broadcast Internet control-message-protocol echoes and directs all echo replies back to the victim, flooding its network link.

A sensor network that based on a hierarchical routing mechanism can use a method similar to the egress filtering in Internet gateways, which can help prevent smurf attacks. By verifying the source addresses, parent routers can verify that all routed packets from below could have been originated legitimately by their children.

4. *Black Holes*

Distance-vector-based protocols[14] provide another easy avenue for an even more effective attack. Nodes advertise zero-cost routes to every other node, forming routing black holes within the network[15]. As their advertisement propagates, the network routes more traffic in their direction. In addition to disrupting message delivery, this causes intense resource contention around the malicious node as neighbors compete for limited bandwidth. These neighbors may themselves be exhausted pre-maturely, causing a hole or partition in the network.

8.2.1.4 Transport layer attacks and countermeasures

Transport layer manages end-to-end connections and this layer is needed when the sensor network intends to be accessed through the Internet. The service the layer provides can be as simple as an unreliable area-to-area any cast, or as complex and costly as a reliable sequenced-multicast byte stream. Sensor networks tend to use simple protocols to minimize the communication overhead of acknowledgments and retransmissions. The transport layer can be attacked via flooding or desynchronization.

1. *Flooding*

The aim of flooding attacks is to exhaust memory resources of a victim system. Similar to TCP SYN flood[16], the attacker sends many connection establishment requests, forcing the victim to allocate memory in order to maintain the state for each connection. Limiting the number of connections prevents complete depletion of resources, which would interfere with all other processes of the victims. However, because the queues and the tables fill with abandoned connections, this method prevents legitimate clients from connecting to the victim as well. Connectionless protocols can naturally resist this type of attack a little, but they may not provide adequate transport-level services for the network.

Client puzzles are a typical way of reducing the severity of flooding attacks by asking all client nodes to demonstrate their commitment to the resources they require. The server can easily create and verify the puzzles. While clients are solving the puzzles, the storage of client-specific information is not required. Servers distribute the puzzle, and clients solve and present them. If the clients hope to connect, they must solve and present the puzzle to the server before receiving a connection. Therefore, an attacker must be able to take more calculated resources per unit time to flood the server with effective connections. Under heavy load, the server measure the puzzles, and learn need work of potential clients. This solution is most suitable for combating adversaries that possess the same limitations as sensor nodes. The downside is that legitimate nodes now have to expend extra resources to get connected, but it is less costly than wasting radio transmissions by flooding.

2. *Desynchronization*

Desynchronization can disrupt an existing connection between two end points. In this attack, the adversary forges messages between endpoints. These messages carry sequence numbers or control flags that lead to the end points request retransmission of missed frames. If the adversary can maintain proper timing, it can hinder the end points from exchanging messages as they will be continually requesting retransmission of previous erroneous messages. Also, this attack leads to an infinite cycle that wastes energy.

This threat is typically countered by authenticating all packets exchanged, including all control fields in the transport protocol header. And then the end points can detect and ignore the malicious packets, assuming the adversary fails to forge the authentication mechanism.

8.2.2 Security requirements

8.2.2.1 Security goals

Various security requirements on sensor networks are presented in almost all the related papers [17–19]. These requirements can be classified into three levels.

1. *Message-based level*

Similar with that in conventional networks, this level deals with data confidentiality, authentication, integrity and freshness. Symmetric key cryptography and message authentication codes are necessary security primitives to support information flow security. Also data freshness is necessarily required as lots of content-correlative information is transmitted on a sensor network during a specific time.

2. *Node-based level*

Situations such as node compromise or capture are investigated on this level. In case that a node is compromised, loaded secret information may be improperly used by adversaries.

3. *Network-based level*

At this level, more network-related issues are addressed, as well as security itself. A major benefit of sensor networks is that they perform in-network processing to reduce large streams of raw data into useful aggregated information. Protecting it is critical. The security issue becomes more challenging when discussed seriously in specific network environments. Firstly, securing a single sensor is completely different from securing the entire network, thus the network-based anti-intrusion abilities have to be estimated. Moreover, such network parameters as routing, node's energy consumption, signal range, network density and etc., should be discussed correlatively. Moreover, the scalability issue is also important with respect to the redeployment of node addition and revocation.

8.2.2.2 Performance Metrics

As addressed above, it's definitely insufficient to access a scheme based on its ability to provide secrecy. Reference [19] proposes the following evaluation metrics.

(1) *Resilience against node capture.* On the network-based level, the fraction of total communications that are compromised is required to be estimated once a capture of several nodes occurs.

(2) *Resistance against node replication.* This issue needs to be seriously investigated as the captured node may be cloned and thus adversaries gain more control of the network.

(3) *Revocation.* Like regular process on node addition, the revocation mechanism is always necessary for detection and insulation of the misbehaving nodes.

(4) *Scale.* Performance of the above security characteristics needs to be generally inspected, corresponding to different network scales.

8.2.3 Hierarchical Architecture for WSN Security

8.2.3.1 Three-level security requirements architecture on security mechanisms

In order to give a general view on security issues addressed in sensor networks, we present the security architecture of sensor networks in Fig. 8.1. As described above, three-level security requirements outline the principles of algorithm design on security mechanisms. We list the corresponding issues for each level in detail. In order to achieve securing available communications and applications in sensor networks, such as identity authentication, routing, data aggregation and etc., most security research focuses on the following three aspects: security primitives, key management and network-related security strategies. Security primitives manage a minimal protection to information flow and a foundation to create secure protocols. Those security primitives are systematical key encryption (SKE), message authentication codes (MAC), and public key cryptography (PKC). The issue of network-related security strategies combines communications throughout the entire network, integrates power and routing awareness, and promotes holistic working performance within tolerable costs[20].

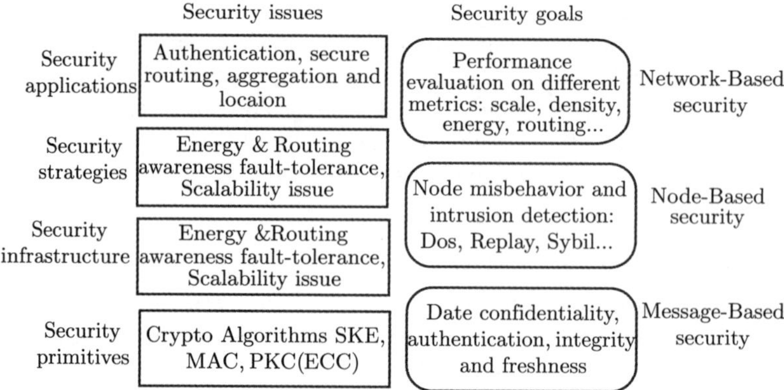

Fig. 8.1 Security architecture of sensor networks.

8.2.3.2 Security architecture (security map) of security issues in WSN

The new security architecture (security map) of security issues in WSN is drawn as in the following Fig. 8.2. Security must be justified and ensured

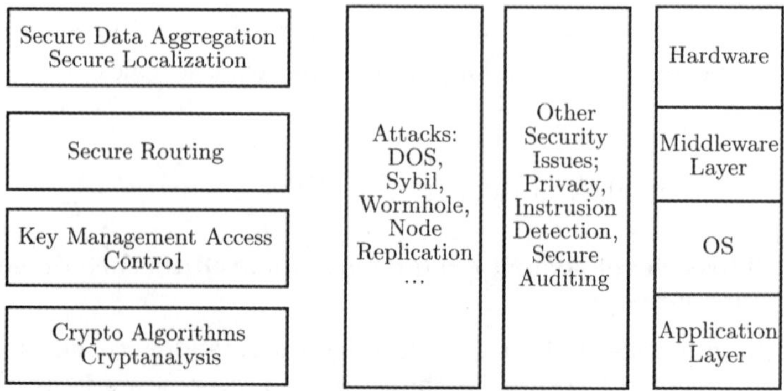

Fig. 8.2 Security Architecture security issues in WSN.

before the large scale deployment of sensors. The vertical comparison in Fig. 8.2 shows that various security issues are rendered in every layer of the protocol stacks from physical layer to application layer. Although it is extremely hard to guarantee the security of every layer, we can deal with the problems one by one and build appropriate security mechanisms satisfying particular appliances.

8.3 Cryptographic Approaches

In WSNs, four major security requirements are integrity, confidentiality, authentication, and freshness. To prevent the network from being attacked, a security scheme should be capable of protecting each data packet within the network from being eavesdropped (confidentiality), altered (integrity), spoofed (authentication), and replayed (freshness). Encryption is used to ensure the confidentiality. A message authentication code (MAC), functioning as a secure checksum, provides the data integrity and authentication in the network.

Symmetric key ciphers and asymmetric key ciphers are the two fundamental categories of ciphers. The security of asymmetric cryptography depends on the difficulty of a mathematical problem and the resulting algorithm consumes considerably more energy than symmetric key ciphers, which are constructed by iteratively applying simple cryptographic operations. Hence in WSNs, the symmetric key cipher is typically utilized to encrypt data during the transmission of sensor data, conforming to the limited energy source in the sensor device[21].

8.3.1 Communication secrecy

Perrig et al.[17] presents a suite of security protocols optimized for sensor networks: SPINS. SPINS consists of two secure building blocks: SNEP (Sensor Network Encryption Protocol) and μTESLA. The function of SNEP is to provide data confidentiality, two-party data authentication, and evidence of data freshness. μTESLA provides authenticated broadcast for severely resource-constrained environments.

1. *SNEP*

SNEP shows a lot of unique advantages. Firstly, it has low communication overhead, which only adds 8 bytes per message. Secondly, it uses a counter, but we avoid transmitting the counter value by keeping state at both end points. Thirdly, SNEP achieves semantic security, a strong security property which prevents eavesdroppers from inferring the message content from the encrypted message. Finally, the same simple and efficient protocol also gives us data authentication, replay protection, and weak message freshness.

Data confidentiality: data confidentiality is one of the most basic security primitives and it is used in almost every security protocol. A simple form of confidentiality can be achieved through encryption, but pure encryption is not sufficient. Semantic security is another significant security property, which ensures that an eavesdropper has no information about the plaintext, even if it sees multiple encryptions of the same plaintext[22].

Two-party authentication and data integrity: it uses a message authentication code (MAC) to achieve two-party authentication and data integrity. A good security design practice is not to reuse the equally cryptographic key for different cryptographic primitives, which hinders any potential interaction between the primitives that might introduce a weakness. Hence we derive independent keys for encryption and MAC operations. The two communicating parties A and B share a master secret key x_{AB}, and they derive independent keys using the pseudorandom function F: encryption keys $K_{AB} = F_x$ (1) and $K_{BA} = F_x$ (3) for each direction of communication, and MAC keys $K'_{AB} = F_x$ (2) and $K'_{BA} = F_x$ (4) for each direction of communication .

The encrypted data has the following format: $E = \{D\}_{\langle K,C \rangle}$, where D is the data, the encryption key is K, and the counter is C. The MAC is

$$M = MAC(K', C||E).$$

The complete message that A sends to B is $A \to B$.

2. μ*TESLA*

Reference [17] makes contributions on providing the authentication scheme (μTESLA)[17,23] through a delayed disclosure of symmetric keys BS-to-all nodes communications. The authors first create a key chain $K_0, K_1, K_2, \cdots$, and the key K_0 (or K_B) is loaded in every node before deployment. Except K_0, each key of the key chains corresponds to a time interval and all packets sent within one time interval are authenticated with the same key. μTESLA achieves authenticated broadcast by two steps: The sender first broadcast the packets along with their MAC. Since the message is encrypted with K_i at that time, no one does know if that message is not a spoof from an adversary. After a time interval δ, the sender then broadcasts the key K_i. By verifying $K_0 = h^i(K_i)$, the receiver then authenticates the packets received at a time interval δ before it is actually broadcasted by the sender. However, μTESLA is designed for base station broadcast. It is much more complicated when this issue is addressed in node-based broadcast.

8.3.2 Achievements on node authentication

Whilst allowing for detection, node authentication can also prevent most of the damage that can be done by malicious intruders. Authentication is a mechanism whereby the identity of a node in a network can be identified as a valid member of the network and as such data authenticity can be achieved. This is where the data is appended with a message authentication code (MAC) and can only be viewed by valid nodes capable of decrypting the MAC, through some determinable means. Any messages received from unauthorized network users can be discarded. There are a number of methods to achieve authentication. These range from device-to-device protocols. However, the authentication include the two killer aspects — entity authentication, and message authentication.

1. *Distinguishes between message and entity authentication*

There are two differences between message authentication (data-origin authentication) and entity authentication. Firstly, message authentication does not provide timeliness guarantees as to when it was created etc., while in entity authentication, time is important, as in this protocol corroboration of a claimant's identity takes place. Secondly, message authentication simply authenticates one message; the process needs to be repeated for each new message. Entity authentication authenticates the claimant for the entire duration of a session.

2. *Data authcntication*

For many applications in sensor networks (including administrative tasks such as controlling sensor node duty cycle or network reprogramming), message authentication is very important. Since an adversary can easily inject message, the receiver needs to ensure that data used in any decision-making process originates from a trusted source. Informally, data authentication allows a receiver to verify that the data really was sent by the claimed sender. In the two-party communication case, data authentication can be achieved through a purely symmetric mechanism: The sender and the receiver share a secret key to compute a message authentication code (MAC) of all communicated data. When a message with a correct MAC arrives, the receiver knows that it must have been sent by the sender.

However, without placing much stronger trust assumptions on the network nodes, this authentication style cannot be applied to a broadcast setting. When a sender sends authentic data to mutually distrusted receivers, it is insecure that using a symmetric MAC because any receivers know the MAC key and could impersonate the sender and forge messages to other receivers. Therefore an asymmetric mechanism is also needed to achieve authenticated broadcast. [17]

3. *Entity authentication*

This authentication is designed to let one party prove the identity of the other. An entity can be a person, a process, a client, or a server. Proving of entity identity needs to be known as the claimant, trying to prove the identity of the claimant party is called the verifier. Typically, base stations or users issue kinds of tasks commands to nodes; then nodes start to work accordingly, gathering data and transmitting to base stations or users. In order to function properly, users and base stations should be authenticated to be the acclaimed entities by nodes. This is because, without entity authentication, adversaries can easily abuse the sensor networks to collect information maliciously or launch energy-exhaustion denial-of-service attacks by frequently ordering nodes to perform nonsense tasks. On the other side, nodes should also be authenticated by base station, other nodes, and users. Otherwise, adversaries can corrupt the result of information collection by inserting invalid nodes into sensor networks. Moreover, any further advanced access control mechanisms require entity authentication[24].

- Basis of entity authentication.[24]
  - Something known: this category includes standard password, PIN (personal identification numbers), etc.
  - Something possesses: they include hand-held customized calculators, magnetic-striped cards etc.
  - Something inherent: examples characteristics like finger prints, handwritten signatures, voice, i.e. some human physical characteristic.
- Types of entity authentication protocols.
  - Weak authentication. This is one of the most conventional schemes where a user has a user id and a password. User id acts like a claim and password as evidence supporting the claim. The system checks to see if it matches or not. Here demonstration of knowledge of the secret which is password in this case; corroborates that the person is verified.
  - Towards strong authentication. Let H be a one-way function. User A begins with secret w. A sends $w_o = H^t(w)$. B initializes its counter for A to i_A =1. The ith identification proceeds from $A \rightarrow B : A, i$, $w_i(H^{t-1}(w))$. B checks that $i = i_A$ and that the received password w_i satisfies $H(w_i) = w_i - 1$. Once verified and successful it sets $i_A = i_A + 1$ and saves w_i.
  - Strong authentication. The basic idea of this authentication is that one entity "proves" its identity to another entity by demonstrating knowledge of a secret known to be associated with that entity, without revealing the secret itself to the verifier during the authentication process.
  - Zero authentication. To address the impersonation issues, zero knowledge protocols are used. It allows a claimant to demonstrate knowledge of a secret while revealing no information of use to verifier. This protocol involves 3 messages. $A \rightarrow B : cert_A, x = \beta^r \bmod p$: B checks to see that $S(I_A)$, $S(v)$is equal to the value of I_A and v sent in certificate when signed and in return sends "e" $A \leftarrow B$: e(where $1 \leqslant \mathrm{e} \leqslant 2^t$):$A$ checks that the value of e send is in the appropriate range. $A \rightarrow B : y = a\mathrm{e} + r \bmod q$:$B$ now computes $z = \beta^y v^{\mathrm{e}} \bmod p$ and accepts A, if $z = x$.

8.3.3 Approaches on Asymmetric Cryptographic Algorithms Utilization

Public-key cryptography[25] is a form of cryptography in which each user or the device taking part in the communication have a pair of keys, a public key and a private key, and a set of operations associated with the keys to do the cryptographic operations. This cryptographic approach involves the use of asymmetric key algorithms hence it is also known as asymmetric cryptography. Participants who receives messages in such a system first creates both a public key and an associated private key, and publishes the public key. When someone wants to send a secure message to the creator of these keys, the sender encrypts it (transforms it to secure form) using the intended recipient's public key; to decrypt the message, the recipient uses the private key.

According to the above discussing, unlike symmetric key algorithms, a public key algorithm does not require a secure initial exchange of one or more secret keys between the sender and receiver. The particular algorithm used for encrypting and decrypting was designed in such a way that, while it is easy for the intended recipient to generate the public and private keys and to decrypt the message using the private key. And it is very difficult for anyone to figure out the private key based upon their knowledge of the public key, while it is easy for the sender to encrypt the message using the public key.

- Asymmetric encryption algorithms.
  - RSA is the most popular asymmetric algorithm that is used for Encryption, Signature and Key Agreement. RSA uses public and private keys that are functions of a pair of large prime numbers. The difficulty of factoring large integers determines its security. In RSA algorithm, the keys are generated by using random data and used for encryption and decryption. The key used for encryption is a public key and the key used for decryption is a private key. Public keys are stored anywhere publicly accessible. The sender encrypts the data using public key, and the receiver decrypts it using his/ her own private key. In that way, no one else can intercept the data except receiver.
  - The Digital Signature Algorithm (DSA) is a public key algorithm that is used for Digital Signature. The DSA standard is specified FIPS 182-2, Digital Signature Standard. It was proposed by the National Institute of Standards and Technology (NIST) in 1991.
  - Pretty Good Privacy (PGP) is a public-private key cryptography system. It allows for users to integrate the encryption's use more easily in their daily tasks, such as e-mail protection and authentication, and protecting files stored on a computer. PGP was originally designed by Phil Zimmerman. It uses IDEA, CAST or Triple DES for actual data encryption and RSA (with up to 2048-bit key) or DH/DSS (with 1024-bit signature key and 4096-bit encryption key) for key manage-

ment and digital signatures. The RSA or DH public key is used to encrypt the IDEA secret key as part of the message.

- Symmetric vs. asymmetric encryption algorithms. Symmetric encryption algorithms encrypt and decrypt with the same key. Main advantages of symmetric algorithms are its security and high speed. Asymmetric encryption algorithms encrypt and decrypt with different keys. Data is encrypted with a public key, and decrypted with a private key. Asymmetric encryption algorithms are incredibly slow and it is impractical to use them to encrypt large amounts of data. Generally, symmetric encryption algorithms are much faster to execute on a computer than asymmetric ones. In practice they are often used together, so that a public-key algorithm is used to encrypt a randomly generated encryption key, and the random key is used to encrypt the actual message using a symmetric algorithm.
- The two main branches of asymmetric encryption algorithms.
  - Public key encryption: it is presumably that anyone cannot decrypt a message encrypted with a recipient's public key except a possessor of the matching private key, this will be the owner of that key and the person had access to the public key used. This is used for confidentiality.
  - Digital signatures: Using digital signature a message can be signed by a device using its private key to ensure authenticity of the message. Any device that has got the access to the public key of the signed device can verify the signature. Therefore, the device receiving the message can ensure that the message is indeed signed by the intended device and is not modified during the transit. And the signature verification would fail, if any the data or signature is modified. A digital signature scheme typically consists of three algorithms: key generation algorithm, signing algorithm and signature verifying algorithm.
- Development.
  - PKC issue in sensor networks has long been considered as "not possible" due to hardware constraints of sensors. However, there is almost no quantitative analysis that supports this widely accepted conclusion. To the best of our knowledge, the first attempt on feasibility of PKC[26] utilization in sensor network environment is reference [27], which is based on available network production ZigBee[28]. In such a network, a new entity called security manager is involved, whose hardware resources are sufficient for public-key operations. The authors of reference [27] propose a hybrid authentication key establishment scheme based on elliptic curve cryptography (ECC)[29]. The introduction of elliptic curve cryptography by Neal Koblitz and Victor Miller independently and simultaneously in the mid-1980s has yielded new public-key algorithms based on the discrete logarithm problem[30]. Mathematically more complex, elliptic curves provide smaller key sizes and faster operations for equivalent estimated security. The scheme puts the cryptographic burden on security manager,

eliminates high-cost public-key operations at sensor side, thus achieves authentication between a sensor and a security manager during key establishment.

- However, in the hybrid scheme sensors are also assumed unable to perform PKC operations. Reference [31] presents the implementation of ECC over F2p for sensor networks based on MICA2[32] mote. Related figures show that public keys can be generated within 34 seconds, and the distribution among nodes of shared secrets is also achieved within reasonable costs. The latest research[33] begins to focus on optimization of the essential operations in PKC such as public key authentication. As symmetric-key based protocols are complicated and always subject to attack by adversaries, PKC utilization would be the next research focus in sensor networks security along with preliminary achievements on development of the related productions.

8.4 Resilience on Key Management

Key management plays a very key role in deploying security strategies of sensor networks. Key management is the provisions made in a cryptography system design that are related to generation, exchange, storage, safeguarding, use, vetting, and replacement of keys[34].According to the schemes used to distribute initial keys and the approaches used to negotiate between nodes, key management mechanisms in WSNs can be roughly classified into three categories: centralized schemes, distributed schemes and hierarchical schemes. The following introduction focuses on a typical distributed scheme.

8.4.1 Schemes of Key Pre-distribution

The Key Pre-distribution Scheme (KPS)[35] is a most typical distributed scheme, where key information is distributed among all sensor nodes prior to deployment. If knowing which nodes are more likely to stay in the same neighborhood before deployment, keys can be decided a priori. However, because of the randomness of the deployment, knowing the set of neighbors deterministically might not be feasible[36]. Current research pays more and more attentions on practical pairwise key pre-distribution scheme, which enables any two sensors to communicate securely with each other.

1. *Probabilistic key distribution*

Probabilistic key distribution scheme is designed to make sure that at least a key-shared path exists in "almost certain" situation. Reference [35] presents the idea of probabilistic key-sharing and related shared-key discovery protocol, which makes an important contribution on that kind of algorithm design.

This scheme picks a random pool (set) of keys S out of the total possible key space. For each node, m keys are randomly selected from the key pool S and stored into the node's memory. This set of m keys is called the node's key ring. The number of keys in the key pool $|S|$ is chosen such that two random subsets of size m in S will share at least one key with some probability p.

Reference [19] makes improvements on security strength, which requires q common keys ($q>1$) instead of just one. The composite K takes the form of $K=\text{hash}(k_1||k_2||\cdots||k_q)$. After intensive study, it shows to a remarkable conclusion that the resilience of the network against node capture will be increase due to the increase of the amount of key overlap.

2. *Polynomial pool-based pairwise key predistribution*

A bivariate t-degree polynomial is used to generate keys, but this polynomial-based key pre-distribution scheme can only tolerate no more than t compromised nodes, and the value of t is limited due to the memory constraints of sensor nodes[37]. The idea of a pool of multiple random bivariate polynomials is desirable. The basic idea of the polynomial pool-based scheme can be considered as the expansion on the meaning of "key". In other words, this scheme is also based on the concept of "key pool", whereas keys are expressed as different polynomials. Reference [38] presents an instantiation on this idea, modeling a sensor network with a total of N sensor nodes as an n-dimensional hypercube.

3. *Multiple-space key pre-distribution scheme*

Blom's scheme achieves optimal resilience at the expense of relatively large memory requirement. However, it is vulnerable to preset key for each node by using a generator matrix. Reference [39] presents a multiple-space key pre-distribution scheme base on Blom's scheme. This scheme achieves good—which offers the advantage of requiring much lower memory usage although not optimal resilience. What's more, reference [39] uses the theory of random diagram analyzes the possibility of constructing key connected graph. Furthermore, reference [39] analyzes the relationship between ω and τ, e.g., $\tau \geqslant \sqrt{\ln \frac{1}{1-p_{\text{actual}}}}\sqrt{\omega}$ where $p_{\text{actual}} = 1 - \frac{[(\omega-\tau)!]^2}{(\omega-2\tau)!\omega!}$.

8.4.2 Malicious behaviors analysis on key management

A wireless sensor network, being a collection of tiny sensor nodes with limited resources (limited coverage, low power, smaller memory sizes and low bandwidth), proves to be a viable solution to many challenging civil and military applications. Their deployment, sometimes in hostile environments, can be dangerously perturbed by any type of sensor failure or, more harmful, by malicious attacks from an opponent[40].

8.4.2.1 Node & Key Compromises on random key predistribution scheme

1. *q-composite random key predistribution scheme*

- Description of the scheme. In the basic scheme and q-composite keys scheme, there is no capability for node-to-node authentication. All that any given node A knows about a given neighbor B is that A and B share some set of common keys. There is no concept of a unique identity for B. This is because there is no limit to the number of times a key could be picked for various key rings in different nodes. A scheme called random pairwise scheme[19] is proposed to address this drawback. The scheme has the following properties: perfect resilience against node capture, node-to-node identity authentication, distributed node revocation without base stations, resistance to node replication and generation and comparable maximum supportable network sizes vs. other schemes without authentication.
- Resilience against node capture in q-composite keys schemes. The q-composite key scheme strengthens the network's resilience against node capture when the number of nodes captured is low. Fig. 8.3 shows the fraction of additional communications (i.e., external communications in the network independent of the captured nodes) that an adversary can compromise based on the information retrieved from x number of captured nodes. It is thus immediately clear that the schemes are not infinitely

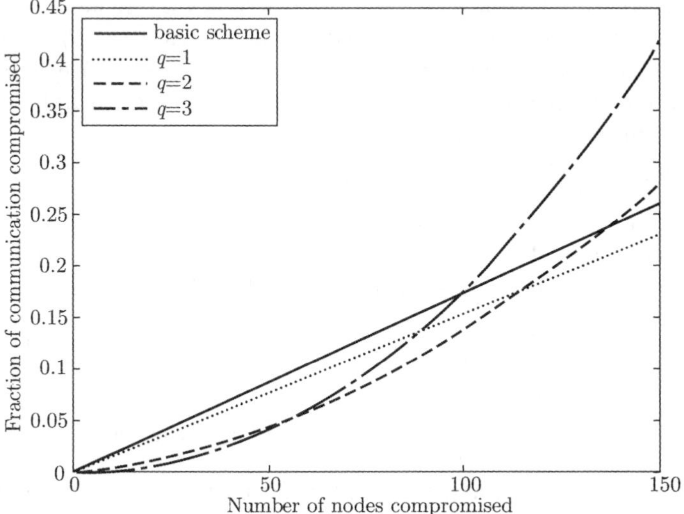

Fig. 8.3 Probability that a specific random communication link between two random nodes A, B can be decrypted by the adversary when the adversary has captured some set of x nodes that does not include A or B. The number of keys stored in each node m=200; the probability of any two neighbors being able to set up a secure link p=0.33.

scalable — a compromise of x number of nodes will always reveal y fraction of the total communications in the network regardless of how large the network is.

2. *Random-pairwise keys scheme*

- Description of the scheme. A new key establishment protocol called the random pairwise scheme is also proposed in reference [19], which has two critical properties of resistance to node replication and generation and perfect resilience against node capture.
- Perfect resilience versus node capture. Because of each pairwise key is unique, capture of any node does not allow the adversary to decrypt any additional communications in the network besides the ones that the compromised node is directly involved in.
- Resistance to revocation attack of distributed scheme. If resistance against node replication is implemented, then the theoretical number of nodes an attacker can revoke per successful node captured is O (d). Because $d = O(\lg n)$, the effectiveness of revocation attack scales only slowly with lg n as network size n increases. Therefore, it is unlikely that an attacker would find it economically worthwhile to launch a revocation attack on the network, especially considering that they must physically establish communications with every node that they wish to revoke[19].

8.4.2.2 Node and key compromise on multiple-space key pre-distribution scheme

The evaluation of multiple-space key pre-distribution scheme in terms of its resilience against node capture is based on two metrics: (a) Probability that at least one key space is broken if x nodes are captured. (b) Fraction of the additional communication (i.e., communication among uncaptured nodes) becomes compromised when x nodes are captured.

1. *Probability of At Least One Space Being Broken*

Firstly, define the unit of memory as the size of a secret key (e.g., 64 bits). Secondly, note that the memory usage is m and each node needs to carry τ spaces. In addition, the value of λ should be $\lfloor \frac{m}{\tau} \rfloor - 1$. By analyzing, we finally get a result that P_r (at least one space is broken $|C_x) \leqslant \omega \cdot \sum_{j=\lambda+1}^{x} \begin{pmatrix} x \\ j \end{pmatrix} \theta^j (1-\theta)^{x-j} = \omega \cdot \sum_{j=\lambda+1}^{x} \begin{pmatrix} x \\ j \end{pmatrix} \left(\frac{\tau}{\omega}\right)^j \left(1-\frac{\tau}{\omega}\right)^{x-j}$, where S_i is the event that space S_i is broken (for $i = 1, \cdots, w$) and C_x is the event that x nodes are compromised in the network.

2. *Fraction of Compromised Network Communication*

Let c be a link in the key-sharing graph between two uncompromised nodes, and let K be the communication key used for this link. Let S_i denote the i_{th} key space, and let B_i represent the joint event that K belongs to S_i and S_i is compromised. Use the notation $K \in S_i$ to represent that "key K was derived using S_i". The probability of c being compromised given the compromise of x other nodes is $\sum_{j=\lambda+1}^{x} \binom{x}{j} \left(\frac{\tau}{\omega}\right)^j \left(1-\frac{\tau}{\omega}\right)^{x-j}$.

Assume that there are γ secure communication links that do not involve any of the x compromised nodes. Given the probability P_r (c is broken $|C_x$), the expected fraction of broken communication links among those γ links is $[\gamma \cdot P_r$ (c is broken $|C_x)]/\gamma = P_r$ (c is broken $|C_x) = P_r$ (S_1 is compromised $|C_x$).

3. *Comparison to previous work*

Figure 8.4 compares the multiple-space key pre-distribution scheme with the Eschenauer-Gligor scheme ($q = 1$) and the Chan-Perrig-Song scheme ($q = 2, 3$). Fig. 8.4 shows that the adversary needs to compromise less than 100 nodes in order to compromise 10% of the links in both the Chan-Perrig-Song scheme and Eschenauer-Gligor scheme, while an adversary needs to compromise 500 nodes before compromising 10% of the links in the multiple-space key pre-distribution scheme. Therefore, this scheme quite substantially lowers the initial payoff to an adversary for small-scale network breaches.

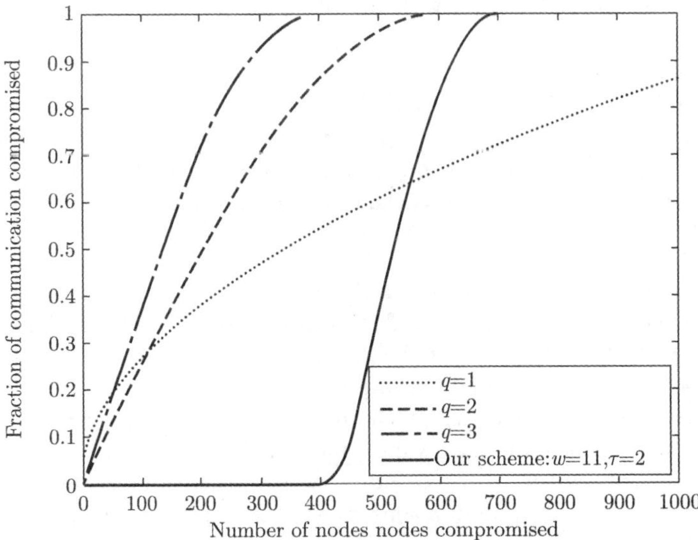

Fig. 8.4 Fraction of compromised links (in the key-sharing graph) between non-compromised nodes, after an adversary has compromised x random nodes. Here, the memory usage of the scheme m=200 and the probability that any given pair of nodes can directly establish a pairwise key $p_{actual} = 0.33$.

In Figure 8.4, it is considered the security performance of the multiple-space key pre-distribution scheme when two neighboring nodes can directly compute a shared key. Since the local connection probability is less than 1, two neighboring nodes might need to use a multi-hop path to set up a shared key. It refers to the secure channel established in this way as an indirect link. When any node or link along the multi-hop path used to establish an indirect link is compromised, the indirect link itself is also compromised.

8.5 Trust Management

Traditional cryptography-based security mechanisms can resist external attack, but can't solve internal attack effectively that was caused by the easily captured nodes[41]. Trust management has now become an additional means to cryptography-based security measures, which can identify selfish and malicious nodes efficiently and solve the security problems for node failure or capture in WSNs. Trust management also can deal with this problem efficiently and enhance the security, reliability and impartiality of the system. Many protocols[42–60] address trust management methods in self-organization networks from different views.

8.5.1 Analysis on Node Vulnerabilities

For wireless sensor networks, many factors, such as mutual interference of wireless links, battlefield applications and nodes exposed to the environment without good physical protection, result in the sensor and nodes exposed to the environment without good physical protection, result in the sensor nodes being more vulnerable to be attacked and compromised[61].

1. *Energy constraints*

Energy is perhaps the greatest constraint to sensor node capabilities. Assume that once sensor nodes are deployed in a sensor network, they cannot be recharged. Therefore, the battery charge taken with them to the field must be conserved to extend the life of the individual sensor node and the entire sensor network. Various mechanisms within the network architecture, including the sensor node hardware, take this limitation into account. When applying security within a sensor node, we are interested in the impact that security has on the lifespan of a sensor. The extra power consumed by sensor nodes due to security is related to the processing required for security functions (e.g., encryption, decryption, signing data, verifying signatures), the energy required to transmit the security related data or overhead, and the energy required to store security parameters in a secure manner (e.g., cryptographic key storage). Since the amount of additional energy consumed for protecting each message is relatively small, the greatest consumer of energy in the security realm is key establishment[62].

2. *Inability of tamper resistance*

As sensor nodes may be deployed in hostile or unattended areas, they would take much risk of physical attack by an adversary. In the worst case, sensible information stored in a sensor node may be compromised, causing some part of the network vulnerable to security attack.

3. *Hardware constraints*

References [17,18] provide detailed performance parameters for prototype of their own productions. For example, Smart Dust nodes are equipped with 8-bit processor, 512 bytes RAM, and 8 Kbytes flash memory for instructions execution. Only 4,500 bytes are available for application code space. Although hardware performance has improved greatly according to the latest figures offered by reference [64], the available resources of sensor nodes are still very tight.

4. *Selfish node*

In the first type, the packet forwarding function performed in the selfish node is disabled for all packets that have a source address or a destination address different from the current selfish node. However, selfish node participates in the route discovery and route maintenance phases of the on-demand protocol. The type 2 model selfish nodes do not participate in the route discovery phase of the reactive protocol. The impact of this model on the network maintenance and operation is more significant than the first one. A selfish node of this type uses the node energy only for its own communications.

8.5.2 Detection schemes on malicious nodes

Other related work lies in the area of misbehavior identification and isolation. Following is a brief discussion of related methods on misbehavior identification.

8.5.2.1 Method-based detection strategies

1. *Local monitoring based detections*

Watchdog mechanism proposed in references [65] is a monitoring method used widely in ad hoc and sensor networks. It is the base of a majority of misbehavior detection algorithms and trust or reputation systems as well. Watchdog detects misbehaving nodes by overhearing transmission. It maintains a buffer of recently sent packets and comparing each overheard packet with the packet in the buffer to see if there is a match. If so, the packet in the buffer is removed and forgotten by the watchdog, since it has been forwarded on. If a packet has remained in the buffer for longer than a certain timeout, then it increases a failure tally for the node responsible for forwarding on the packet. If the tally exceeds a certain threshold bandwidth, it determines that the node is misbehaving and sends a message to the source notifying it of the misbehaving node. Fig. 8.5 illustrates how the watchdog works. Assume there exists a path from node A to D through intermediate nodes B, S, and C. Node B is not able to transmit all the way to node C, but it can listen to node S's traffic. Thus, when B transmits a packet for S to forward to C, B can often tell if S transmits the packet. If encryption is not performed separately for each link, which can be expensive, then B can tell if S has tampered with the payload or the header as well.

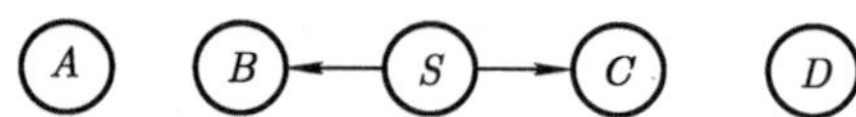

Fig. 8.5 When S forwards a packet from A toward D through C, B can overhear S's transmission and can verify that S has attempted to pass the packet to C. The solid line represents the intended direction of the packet sent by S to C, while the dashed line indicates that B is within transmission range of S and can overhear the packet transfer.

LiteWorp scheme[66] can detect the malicious nodes by local monitoring. If a node finds its neighbor discarding packets or forwarding wrong packets, it will increase the malicious behavior value of this neighbor. When a node finds its neighbor's malicious behavior value exceeds the threshold, it will remove the neighbor from its neighbor list. Reference [67] proposes a mechanism DESCM (Detection and Location of Malicious nodes based on Source Coding and Multi-path transmission), which does not require any other special hardware or the mechanism of encryption and authentication. After determining the path with malicious nodes, DESCM can detect and locate the malicious nodes based on local monitoring or analysis of detection replies. Furthermore, Huang Lei, et al.[3] design an extended watchdog mechanism named last-hop malicious node detection and avoidance (LHDA) algorithm.

2. *Hop count based detections*

EDWA (End-to-end Detection of Wormhole Attack)[68] estimate the number of hops between two nodes according to Euclidean distance estimation model. Then EDWA can detect and locate the malicious nodes by comparing the estimated hops with feedback ones.

There are three steps are involved: source node applies wormhole detection in each route discovery based on the shortest path estimation; once a wormhole is detected, a wormhole tracing phase will be launched by the source to identify the two end points of the wormhole. Then the source selects a shortest path from the legitimate routes set for data communication. EDWA needs special hardware facilities to support. Based on the information of the nodes' neighbor, a central controller can reconstruct the topology of the sensor network using Dijkstra algorithm. The malicious nodes can be located by detecting the bending features on the rebuilt network topology. In this mechanism, the rate of detecting malicious nodes incorrectly will be highly increased when the sensors are deployed in some complex area.

3. *Probing based detections*

References [48,49] propose two schemes based on probing: expanding Time-To-Live search (E-TTL) and Binary Search mechanisms. In E-TTL the sink sends probe packets with increasing hop-count. Each intermediate node decrements the hop-count before forwarding. When the hop count reaches zero at a node, that node sends ACK to the sink informing it of its location and that the packet was received safely. Hence, the sink identifies that part of the path as safe and increases the hop count in subsequent packets. Alternatively the TTL can also be increased exponentially rather than linearly, which gives rise to less delay than basic E-TTL, and may also be restricted to a small number. Binary Search mechanisms probe nodes along a suspected path using inputs from intermediate nodes and an expanding ring probing. This phase discovers faulty links on the path from the source to the destination in O (lg n) probes, where n is the average length of the path. A black list of the malicious hosts is broadcast via trusted neighbors until it reaches the neighbor of that malicious host.

8.5.2.2 Measure-based detection strategies

1. *The statistics-based malicious node detection scheme*

A statistics-based malicious node detection scheme is proposed by Ana Paula R. da Silva etc. in references [69]. In such a scheme, a series of regulations are predefined to describe the normal behaviors of nodes and further judge the anomaly behaviors of nodes. And the rate of false alarming is quite high because there is no interaction among nodes. A similar identification system is proposed by I. Khalil etc. This identification system adds the interactive link among nodes.

2. *The rule-based malicious node detection scheme*

A rule-based malicious node detection scheme in Ad Hoc is proposed by Chin-Yang Tseng etc. in references [70]. This scheme uses the monitoring points distributing in the network to monitor nodes whether operate in accordance with the routing norms in the process of AODV route query phase, then a finite state machine formed by the norms is used to identify nodes as normal state, suspected state, and intrusion state.

8.5.3 Trust Computing

The expected contribution is building a probabilistic framework model to calculate and continuously update trust value between nodes in wireless sensor networks based on the sensed event and to exclude malicious and faulty nodes from the network. In other words, creating a framework to maintain the security and the reliability of a sensor network by examining the trust between nodes, so every node has a trust value for every other node in the surrounding area and based on that value the cooperation occurs between nodes[71].

8.5.3.1 The procedures of trust computing

1. *Trust predefined*

Trust levels can be represented in different schemes such as continuous values in the range of $(-1, +1)$ or discrete values with labels rather than numbers, such as very low trust, low trust, medium trust, high trust, very high trust and blind trust depends on the environment it is implemented in. Trust degrees can be represented as simple values, such as trusted and distrusted or as structured values of at least two elements, where the first element represents an action, say access a file, and the second element represents the trust level associated to that action. Trust levels can also be computed based on the effort that one node is willing to expend for another node. This effort can be in terms of battery consumption, packets forwarded or dropped or any other such parameter that helps to establish a mutual trust level[71].

The benefit of using values for trust is that it reflects the continuous nature of trust in WSN and it allows easy implementation and experimentation. The drawback is that the subjectivity is more difficult to understand and the sensitivity may be a problem because small differences in individual values may produce relatively large differences in the overall result.

2. *Trust value initialization*

Trust value initialization is directly related to trust predefined. All nodes are initialized to the trust of the value of the minimum, maximum value and the middle values.

3. *Synthesis of trust value*

In WSNs, the merger of trust value often uses simple calculate method of addition and ratio, which uses simple calculation model to save energy consumption. Overall, the synthesis of trust value includes transverse synthesis, vertical synthetic and hierarchy synthesis.

Transverse synthesis: LS (local sum) is the sum of local information (LI) of nodes which are evaluated. RS (reputation sum) is based on reputation given by other nodes. LRS (local-reputation sum) is the merger of local information and reputation information.

Vertical synthetic: Vertical synthetic refers to the trust calculation in the direction of time axis, combining by recent trust and past trust value which is also called the updates of trust.

Hierarchy synthesis: Sometimes, the filtering synthesis of trust exists in the hierarchical trust management system.

Analysis of computational models for trust management in WSNs displayed in Table 8.1.

Table 8.1 Analysis of computational models for trust management in WSNs

	Trust management system	Trust factors	Trust evaluation	Transverse integration	Vertical integration	Hierarchy integration	payload
	[72].PLUS [73]	T,C	LS−RS−LRS	−	−		A
	TRANS [48][49]	T,C,O	LS	−	−		MP
Simple	SecCBSN [55]	T,O	RS−TU	V	−		A,MP
Weight	GTMS [74]	I	(LS+TU)−RS−LRS	TR	R	GT	A
Model	RFSN[44]	T,C,D,O	(LS+TU)−(RS+TU)−LRS	TR,PR	R	−	A,MP
	[43]	T	LS	V	−	−	A
	[50][51]	D,O	LS−RS	TR	−	GT	A
Exponential model	TIBFIT [46]	D	TU	−	P	−	−
	[75]	D	TU	−	P	−	−
Statistical model	BRSN[44]	T,C,D,O	LTR+TU	TR,PR	R	−	A,MP
	[76]	T,D	LS−RS−LRS	TR	R	−	MP
Game theory model	[77]	T,O	LS	−	−	−	−

I: Number of success/fail interactions, T: Transmission factors, C: Cryptography factors, D: Application data factors, O: Other factors; TR: Take (functions of)trust values of judges as the coefficients of reputations they sent, PR: Only "good" reputations considered, V: Vote; R: Higher proportion of RT, P: Higher proportion of PT; GT: Group trust value computing; A: Aggregation, MP: Packages specially for trust evaluation (acknowledge packages, beacon, et al.); L: Low

8.5.3.2 Trust formation algorithms

The trust management methods can be classified into two categories: distributive authorization system based on trust chain and network trust evaluation system based on nodes' behaviors[78–81]. In the former system, the authorized individual is allowed to collect all the information of other authorized ones. It checks the consistency through strategy inference engine in light of local policy and authorization requirements. In addition, if a trust chain exists between two strange individuals, the authorization is able to be relayed by signing indirect objects which have trust rights. That is to say, the authorization individual has rights to deal with its trusted objects. But it is very dangerous for the limited resources of WSNs when the authorization nodes are compromised. In the latter system, individuals acquire all kinds of related information, including the actions of evaluated individuals, interacting rules and other individuals' opinions. Then, the sensor nodes obtain other nodes' trust value by different computing method in application. This trust management method has advantages of less resources consumption, peer-to-peer structure and no centers. Therefore, trust management schemes similar to the latter one are more frequently applied in the WSNs.

Most of the definitions of trust in the literature are focusing on what trust is used for in a static fashion and not on the dynamic aspects of trust such as the formation, evolution, revocation and propagation of trust. Trust formation in WSN is the process of establishing the initial trust between nodes. The trust calculation in WSN mainly consists of three parts: communication trust, data trust and energy trust. Communication trust includes direct trust (node's previous experience) and direct trust (recommendations from surroundings nodes). Fig. 8.6 shows a general trust computational model used to calculate trust values in WSN.

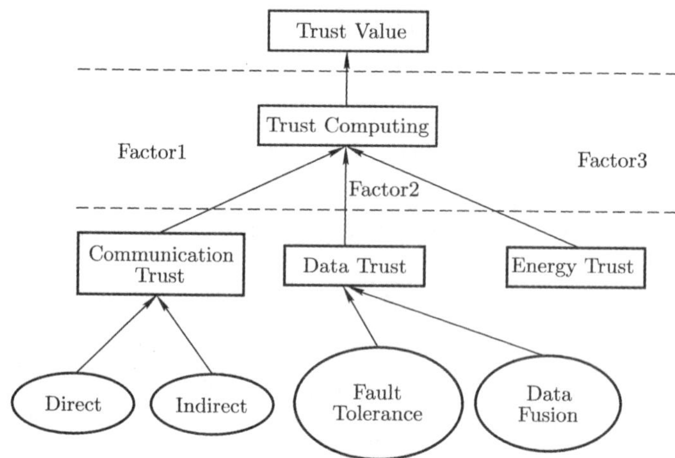

Fig. 8.6 General trust computational model.

In order for nodes in a network to receive updates regarding the trusted behaviors of nodes or even threats, a mechanism for trust reporting is necessary. Calculations of trust levels and trust relationship establishment depend on trust reports.

- *Communication trust computing.* Communication trust means the relationship value calculated between two cooperation nodes in a wireless sensor network which can send or receive information each other. It is a common trust evaluation mechanism that can identify malicious node and selfish node through the observation of communication behavior, which including direct trust and indirect trust. About the trust computing of WSNs, researchers have put forward several ways of calculating the trust value in different application fields. Srivastava and Ganeriwal established the Beta Trust Model for WSNs that was based on the work of Josang and Ismail[82] to acquire the reputation rating of transaction node in the electronic commerce. Srinivasan et al.[83] also mentioned the probability of using of the Beta Reputation System in WSNs.
- *Data trust computing.* Data trust refers to the trust assessment of the fault tolerance and consistency of data. The trust model presented by Josang[84,85] was used to deal with uncertainties of data stream in WSNs.

Krasniewski put forward a fault-tolerant system TIBFIT[46] based on trust in order to compute the trust value of node in WSNs with the structure of cluster. And Hur[50,51] presented a security data fusion algorithm based on trust which calculated the trust value of data fusion by examining the consistency of the data. Reference [41] combine above mentioned methods to develop a simplified method of calculating the data trust value.

- *Energy trust computing.* Energy trust in WSNs refers to the existing energy of node whether lower than a set threshold and whether to complete the new communications and data-processing tasks. If the energy of a node was consumed excessively, the survival period of WSNs would be sharply reduced. Therefore, we can know the existing energy of a node anytime through calculating the energy trust value in order to avoid the low competitiveness nodes excessively used.

8.5.3.3 Trust Routing for Location-aware Sensor Networks (TRANS)

TRANS[48,49] is proposed by Tanachaiwiwat et al, which uses the concept of trust to select a secure path that do not include misbehaving nodes by identifying the insecure locations and routing around them efficiently.

1. *Trust factors and trust value*

- Cryptography (C_i), Sensors supporting cryptography for encryption are given a higher trust value ($C_i = 1$), and are able to authenticate the sink's messages unless compromised.
- Availability (A_i), $A_i = \dfrac{\sum_{j=1}^{n} QA_j}{n}$ where $QA_j = 0$, otherwise $QA_j = 1$.
- Packet forwarding (P_i), $P_i = \dfrac{\sum_{j=1}^{m} QP_j}{m}$ where QP_j represents *jth* reply status; if the request/reply is received then $QP_j = 1$, otherwise $QP_j = 0$.
- $T_i = C_i \cdot A_i \cdot \beta P_i$ (T: Trust value).

2. *Trust routing analysis*

It is assumed that sensors know their (approximate) locations and that geographic routing is used. And assume that all destination nodes use the loose-time synchronization asymmetric mechanism, TESLA, to authenticate all requests and that the shared encryption key will be carried with the authenticated message from sink or base station to ensure message confidentiality. Based on this information, each node initializes trust values for its neighbors' locations. Then, the sensors and sinks monitor the activities of their neighbors and adjust their trust values accordingly. A trusted neighbor is a node that can decrypt the request and has enough trust value (based on forwarding history as recorded by the sink and other intermediate nodes). A sink sends a message only to its trusted neighbors for the destined location. Those neighbors correspondingly forward the packet to their trusted neighbors that have the nearest location to destination. Thus the packet reaches the destination along a path of trusted nodes.

3. *Identifying and isolating insecure location*

The model propose and study several schemes for probing to identify insecure locations, including expanding TTL ring search, binary search and one shot. It also introduces two schemes for isolating insecure locations: black list flooding and embedded black list (or detour points). In the first approach the sink floods the black list to the vicinity of the insecure location. This scheme does not require modification of GPSR routing or to the packet header because the non-cooperative node (at the insecure-location) will be simply removed from the neighbor list and will not be selected to participate in any routing activity. In the second scheme the sink includes the black list information in the header of packet and sends directly to a detour point. This approach incurs less packets overhead but requires modification of packet headers and possible simple extensions to GPSR to route to detour points.

4. *Advantages and disadvantages*

The main contribution of this approach lies in the explicit design and trade-off between secure/trust routing and shortest path routing, the illustration of the route infection problem and the introduction of several node isolation schemes. But there is the possibility that some nodes are misjudged to be malicious because of the abominable channel or compromised nodes. Consequently, it requires a mechanism to allow the nodes in black list to turn into usable nodes again, whereas the model neglects this point.

8.5.3.4 A framework for trust-based cluster head election in wireless sensor networks

The election of a malicious or compromised node as the cluster head is one of the most significant breaches in cluster-based wireless sensor networks. This model introduces a distributed trust-based framework and a mechanism for the election of trustworthy cluster heads. If the cluster head is unbelievable, a new one will be elected in another round to avoid effectively malicious or selfish node to act as cluster head[43].

The trust evaluation matrix consists of several trust evaluation factors as follows.

RF_N: Data Packet Received for forward, Data Packet.

F_N: Forwarded.

DM_N: Data Packet Modified.

AM_N: Data Packet Address Modified.

CRF_N: Control Packet Received for forward.

CF_N: Control Packet Forwarded.

CM_N: Control Packet Modified.

CAM_N: Control Packet Address Modified.

$T_N(X_i)$: Trust Level, a, trust level, denoted by $T_N(X_i) = w_1d_1 + w_2d_2 + w_3d_3 + w_4c_1 + w_5c_2 + w_6c_3 + \gamma$, where w_1 to w_6 are weights and γ is a predetermined constant that is set to equal to the average packet drop rate of the network; d_1, d_2, d_3, and c_1, c_2, c_3 are related to the data packets and control packets respectively.

- *Communication from node to cluster head.* After setup, the cluster heads create a time division multiplexing (TDM) schedule and inform each cluster member. The nodes are actively transmitting or listening for a period of the time and off the remainder. The nodes transmit only at their scheduled time. This allows the nodes to listen to the communication in their respective clusters. It is through this passive listening that the nodes are able to develop trust relationships with their neighbors. Nodes that constantly drop packets or which behave in a selective or selfish manner can be easily detected by their neighbors. Each node stores and maintains a trust table of its neighbors.
- *Trust level storage and distribution.* Every node stores a trust table to record the trust levels of each of its neighbors. Neighbors are confined to those within the broadcast radius of the node. The mechanism does not encourage sharing of trust information among neighbors and the node does not record a trust level for itself. Trust levels are only sent to the cluster head upon request. What's more, this mechanism can reduces the effect of bad mouthing, since trust computation is not based on second hand observation except by the cluster head in the finally when all the votes are counted. Also, because the nodes do not record their own trust level, it is less likely for malicious nodes to upgrade themselves to high trust levels.
- *Advantages and disadvantages.* This trust model decreases the likelihood of malicious or compromised nodes from becoming cluster heads, which is most suitable for wireless sensor networks due to its minimal energy and computational requirement. However, this centralized trust management model increases the network communication payload and the passive trust decision-making slows down the convergent speed of cluster head election.

8.5.3.5 Reputation Based Framework for Sensor Networks (RFSN)

Ganeriwal and Srivastava[44] propose a framework where each sensor node maintains reputation metrics which both represent past behavior of other nodes and are used as an inherent aspect in predicting their future behavior and employ a Bayesian formulation, specially a beta reputation system, for the algorithm steps of reputation representation, updates, integration and trust evolution. A sensor node continuously builds these reputation metrics for other nodes by monitoring their behavior and rating them as being cooperative (expected behavior of the nodes in the network) or non-cooperative (unexpected behavior that is most likely the result of a system fault or node compromise). Then the node uses this reputation to evaluate the trustworthiness of other nodes and the data they provide.

- *Trust table and process.* R_{ij}: Reputation of node j from the perspective of node i; α and β represents magnitude of cooperation and non-cooperation. T_{ij}: As node i's prediction of the expected future behavior of node j. T_{ij} is obtained by taking a statistical expectation of this prediction.
- *Beta reputation system for sensor networks*[45].

$$R_{ij} = Beta(\alpha, \beta) = \frac{\Gamma(\alpha+\beta)}{\Gamma(\alpha)\Gamma(\beta)} x^{\alpha-1}(1-x)^{\beta-1} \forall 0 \leqslant x \leqslant 1, \alpha \geqslant 0, \beta \geqslant 0$$

$$T_{ij} = E[R_{ij}] = E[Beta(\alpha_j, \beta_j)] = \frac{\alpha_j}{\alpha_j + \beta_j}$$

For one thing, Node i update $R_{ij} = Beta(\alpha_j, \beta_j)$ based on r cooperative and non-cooperative observations about j. For another, node i update $R_{ij} = Beta(\alpha_j, \beta_j)$ by receives reputation information about node j through node k. Node i already have prior reputation information about j and k, represented by (α_j, β_j) and (α_k, β_k) respectively.

- *Advantages and disadvantages.* RFSN is able to identify the misbehaving nodes for a variety of fault scenarios. Besides identifying the misbehavior of the nodes, it is able to establish a relative magnitude of each node's misbehavior as compared to other misbehaving or good nodes as well. However, because the lack of prior knowledge about wireless sensor networks, the model's subjective assumptions of prior distribution aggravate the uncertainty of trust. On one hand, RFSN regards the subject fuzziness of trust as the randomness and use pure probability statistic method to assess trustworthiness, which is difficult to obtain prior knowledge from practical application and inevitably result in something unreasonable. On the other hand, this model fails to provide any confidentiality or the authentication of the readings being reported by individual sensor nodes. It also cannot tolerate a much more planned attack that tries to abuse weaknesses in different building blocks of the framework.

8.5.3.6 Trust index based security data fusion

Hur et al.[50,51] divide the network into several grids, which accomplishes secure data integration by crosschecking the consistency of nodes' data and can identify trustworthiness of sensor nodes in order to filter out malicious nodes' deceitful data.

- *The protocol of this trust evaluation model.* The protocol consists of four steps. Firstly, divide sensing areas into some logical grids and assign a unique identification to each grid. Secondly, sensor nodes deployed in each grid verify location information of their neighbor nodes by ECHO protocol. Thirdly, each node evaluates trustworthiness of its neighbor nodes by crosschecking the neighbor nodes' redundant sensing data with its own result. Inconsistent data from malicious or compromised nodes can be detected in this step. Fourthly, special nodes, aggregators, aggregate sensing data from their grids and transmit the computed results to the destination node, sink. Inconsistent data from malicious nodes can be excluded in this step.

- *Trust evaluation and computation.* Trust factor: sensor nodes evaluate trustworthiness of other nodes. Each sensor node has a trust evaluation matrix which stores the trust evaluation factors for its neighbor nodes. The trust evaluation matrix consists of several trust evaluation factors as follows.

  Identification: $ID_i =< GridID, Position_i >$

  $D_{i,j}$: this factor contains distance information between two nodes.

  S_i: sensing communication value of node i.

  $R_i =< sr_i, st_i >$: Sensing result value of node i.

  C_i: consistency value of node i. This factor represents a level of consistency of a node. Based on this factor, it can identify malicious or compromised nodes, and filter out their data in the networks.

  B_i: Battery value of node I; according to the adoption of this battery factor, we can prevent such biased battery exhaustion.

  T_i: Trust value of node i.

  Trust computation: $T_{ij} = \frac{W_1 C_1 + W_2 S_i + W_3 B_i}{\sum_{i=1}^{3} W_i}$

  $T_i = \frac{\sum_{j=1}^{k} (T_j+1) \times T_{ij}}{\sum_{j=1}^{k} (T_j+1)}$, where k means the number of repliers, T_{ij} means a trust value for node i received from node j and $0 < W_i < 1$. $\sum_{i=1}^{3} W_i \neq 0$.

- *Advantages and disadvantages.* This approach is one of the incipient researches on trust evaluation model for wireless sensor networks that can handle and filter out the inconsistent sensing data of the malicious nodes but collusion attacks are not able to be resisted very well.

8.5.3.7 Trust index based fault tolerance for ability data faults in sensor

The goal of the trust index based fault tolerance for ability data faults in sensor (TIBFIT) protocol is to determine whether an event has occurred from analyzing reports from the event neighbors[46].

The main idea of this protocol is as follows, which is introduced in references [46]. To combat failures in the reporting nodes, each node is assigned a trust index (TI), maintained at the cluster head (CH), to indicate its track record in reporting past events correctly. The TI is a real number between zero and one and is initially set to one. And the node's TI will be decreased if each report a node makes that is deemed incorrect by the CH. Similarly, for each report a node makes that is deemed correct by the CH, the node's TI is increased, but not beyond one. Thus correctly functioning nodes will have a TI approaching one while faulty and malicious nodes will have a lower TI. Assume that correct nodes are allowed to make occasional errors due to natural causes. The rate of these errors is denoted the natural error rate (NER). The TI is decremented exponentially. Nodes that make mistakes are penalized more for earlier mistakes, and find it more difficult to regain their previous trust levels. This is considered better than a linear model where a node that lies 50% of the time would still occasionally have the trust index value of one. If a node errs more frequently than its NER its index decreases, while if it errs less frequently then its index increases. An uncompromised node's TI is expected to remain at the same value.

The TI is calculated as $\mathrm{TI} = \mathrm{e}^{-\lambda v}$, where λ is a proportionality constant that is application dependent. A variable v is maintained for each node at the CH.

8.5.4 Inference-based misbehavior detection

In an adversarial environment, various kinds of security attacks become possible if malicious nodes could claim fake locations that are different from where they are physically located. To address these issues, various methods[86−93] are proposed. They provide a set of effective mechanisms to detect and filter out compromised anchors and nodes. Most approaches depend on a few trusted entities (anchors or nodes), requiring at least the majority of these entities are not compromised. Reference [87] proposes a secure localization mechanism, which significantly different from the existing ones. This approach detects the existence of these nodes, termed as phantom nodes, without relying on any trusted entities.

This approach is based on two factors.

Firstly, prevent the phantom nodes from generating consistent ranging (distance) claims to multiple honest nodes. If the locations of neighboring nodes are known a priori, a set of fake, albeit consistent and ranging distances can be easily created by calculating the distances from a fake location to each of its neighbors' location. Therefore, it is important to hide the location information during the phase of ranging. Without the location information of the neighboring nodes, it is difficult for an attacker to generate a set of consistent ranging values (distances) and hence to fake itself into a different physical location. To prevent phantom nodes generating a set of fake, albeit consistent, ranging claims, it should follow two simple design rules: (a) Accepting only ranging claims, not location claims. (b) Hiding the location information during the ranging phase.

Secondly, if the phantom nodes generate a set of inconsistent ranging claims, speculative method was proposed to detect them.

- *Distance measurement phase.* When the consistent ranging claims by phantom nodes are prevented, we can identify the phantom nodes by detecting the inconsistent ranging claims. Each node v measures the distances to neighbors and disseminate these measurements back to its neighbors. For each collected distance, if $\hat{d}_{ij} = \hat{d}_{ji}(\hat{d}_{ij}$: the measured distance to node j by i), it is included in the filtering phase.

- *Filtering phase.* In this phase, a novel speculative procedure can effectively and efficiently filters out phantom nodes. Initially, the node v picks up two neighbors i and j randomly as pivots. (Note that node i and j could be phantom nodes themselves).Using the node v as the origin, the neighbors i and j and three distance information among v, i and j, the local coordinate system is constructed. A graph $G(V, E)$ is used to construct a consistent subset in the node v's coordinate system. The set V contains the node v and its neighbors, and the set E is used to keep the edges between two nodes when the distance information between them maintains consistency. If the difference of $\hat{d}_{ij}$ (the measured distance between i and j) and $\tilde{d}_{ij}$ (the computed distance between i and j) exceed the threshold, the edge between i and j will be exclude in E. The largest connected set V that contains node v is regarded as the largest consistent subset in the speculative plane L. The largest connected set V that contains node v is regarded as the largest consistent (A set of nodes is consistent, if they can be projected on the unique Euclidean plane, keeping the measured distances among themselves.) subset in the speculative plane L. This filtering procedure is done *iter* times (*iter* is a key parameter), and the cluster with the largest size is chosen as a final result.
- *Identifying consistent subset.* In this process, it shows that (a) the largest cluster must consist of only legitimate nodes, (b) we can determine the case where a chosen pivot is, unfortunately, a phantom node, (c) when all the pivots chosen are honest node, the consistent cluster computed and (d) if at least one of pivots is a phantom node, the size of largest cluster is smaller than the one when none of pivots is a phantom node. As an example, Fig. 8.7, Fig. 8.8 and Fig. 8.9 reflect these properties.

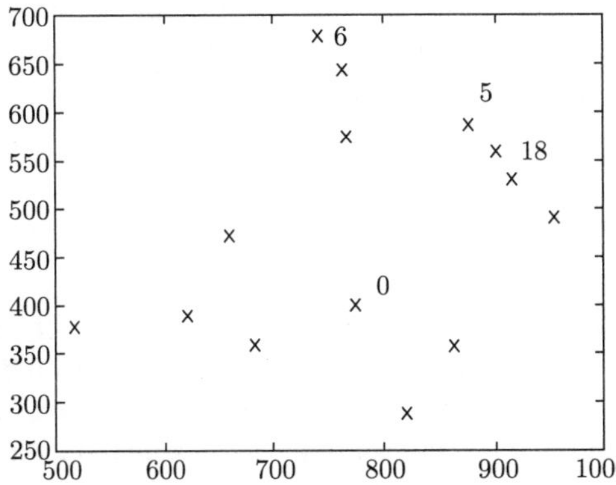

Fig. 8.7 An example plot of actual locations of nodes.

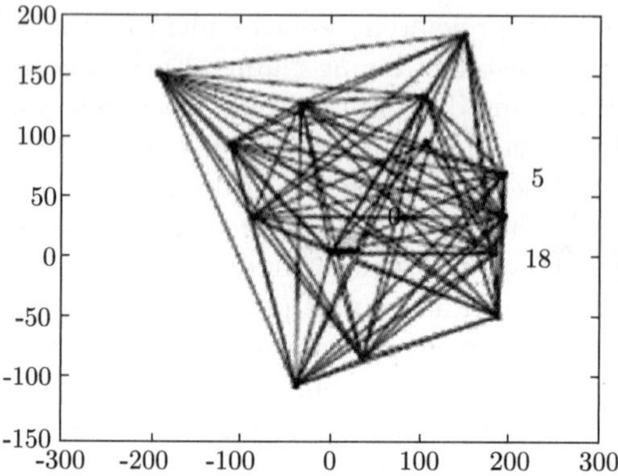

Fig. 8.8 Clusters without phantom pivot.

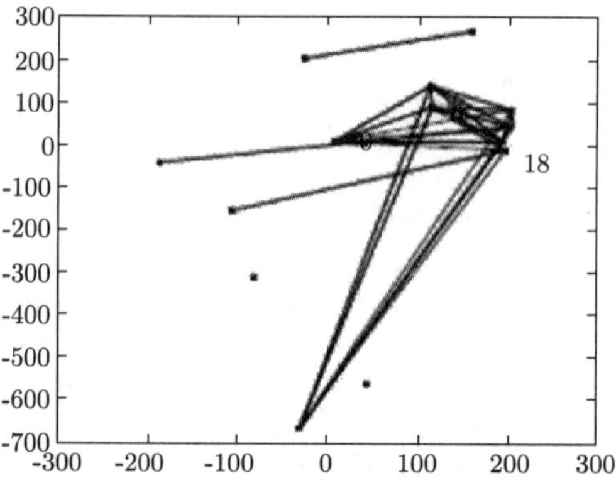

Fig. 8.9 Clusters with phantom pivot.

Figure 8.7 plots the real locations of the nodes, among which node 0 is a verifying node, node 6 is a phantom node, node 5 and 18 are not compromised, Fig. 8.8 shows the cluster created when the pivot is not compromised, Fig. 8.9 is the cluster when the phantom pivot (node 6) is used, whose size is much smaller than the size of cluster.

8.6 Location Privacy

According to different protection objects, the privacy problem in WSNs can be classified into three categories: data privacy, location privacy and identity privacy. A data privacy threat is any means by which an adversary can determine the meaning of a communication exchange. An identity privacy threat is a method that allows an adversary to deduce the identities of entities involved in a communication exchange. Any method that allows an adversary to determine the location of a communicating entity is a threat to that entity's location privacy. These privacy threats are not required to occur together, nor must they occur separately.

In many sensor network applications, location privacy is of particular importance since knowing the locations of data sources and sinks makes it easier to launch various pinpoint attacks. Nevertheless, location privacy protection is a very challenging problem. On one hand, observed events or behaviors of the monitored objects need to be relayed to the access points via multi-hop communication in a sensor network. On the other hand, an adversary can easily track backward and forward along the routing path to identify the data sources and destinations. Currently, a lot of defense strategies have been proposed to protect the location privacy of key nodes in a WSN from being exposed. In addition, location privacy can be classified into four categories: source location privacy, query location privacy, storage location privacy and two-way location privacy. Source location privacy in wireless sensor networks is a very important security issue and we are focus on source location privacy in this chapter.

8.6.1 Flooding mechanisms

The first source node location privacy protection protocol which use flooding for WSNs[97] was proposed by Ozturk et al. They used a metric called safety period to evaluate the performance of a location privacy protocol in the presence of a local attacker. The metric is defined as the number of messages the source node can send before it is localized by the attacker. With this metric, they have evaluated the impacts of three flooding mechanisms on the privacy of source node locations, e.g., baseline flooding, probabilistic flooding, and phantom flooding.

1. *Baseline flooding*

In baseline flooding, every sensor node checks if a received packet is duplicated and rebroadcasts it to all neighbors if it is not, otherwise it discards the duplicated messages. In this mechanism, since all nodes participate in the flooding process, it was believed that the attacker will be effectively misled to wrong source nodes. However, practically the attacker can easily trace to the true source node in this type of flooding. This is because the first packet arrived at the sink node is in fact transmitted along the shortest path between the source node and the sink node; thus the attacker can easily trace the true source node reversely along this shortest path.

2. *Probabilistic flooding*

Probabilistic flooding[98,99] was first proposed as an optimization of the baseline flooding technique to cut down energy consumption. In probabilistic flooding, only a subset of nodes within the entire network participates in data forwarding, while the others simply discard the messages they receive. To address the side effects of baseline flooding, probabilistic flooding is proposed in references [97], in which intermediate sensor nodes forward packets in a probabilistic way. Upon receiving a packet, a sensor node uses a predetermined probability to determine if it should forward the packet. With this method, the route used to deliver the packets from the source node to the sink node are not fixed, which makes it more difficult for the attacker to trace the source node. Nonetheless, it is not guaranteed that all data packets sent by the source node would be received by the base station due to the randomness involved in this approach.

3. *Phantom flooding*

In this flooding scheme, it takes two steps to deliver a packet from the source node to the base station. In the first step, the packet is sent to a random node called phantom node by random walking or direct walking. In the second step, the packet is flooded by the phantom node into the network to reach the base station. The randomness involved in the first step increases the difficulty for the attacker to trace the source node, thus prolongs the safety period. However, with phantom flooding the transmission latency of packets also increases. Fig. 8.10 shows an example scenario of phantom flooding.

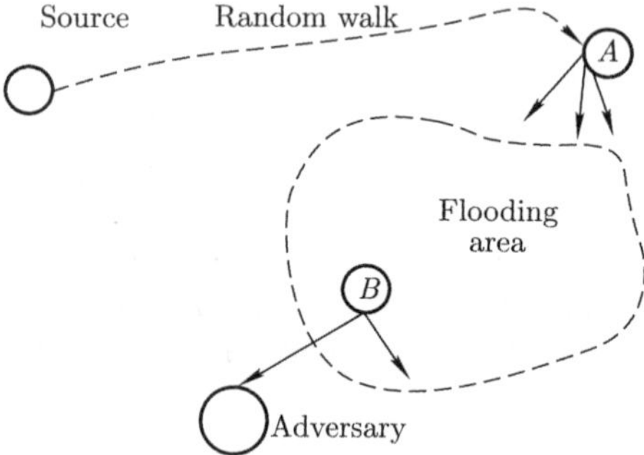

Fig. 8.10 The example scenario of phantom flooding.

Although flooding strategies can help protect the source node location privacy, it is still relatively vulnerable to the hop-by-hop tracing attacks. Furthermore, flooding will consume a large amount of energy in the network and hence may substantially reduce the lifetime of the network.

8.6.2 Random walk strategies

The basic idea of random walk strategies is that every packet takes a different route to the sink node. For every packet sent by the source node, the transmission path is randomly generated therefore not fixed, which increases the length of data transmission paths and decreases the number of packets passing an individual node. With this type of strategies, the attacker may not be able to obtain enough packets to trace the source node successfully. Typical random walk based strategies are described in the following.

1. *Phantom routing techniques*

Kamat P. et al.[94] introduce a new family of flooding and single-path routing protocols for sensor networks, called phantom routing techniques. The goal behind phantom techniques is to entice the hunter away from the source towards a phantom source.

In phantom routing, the delivery of every message experiences two phases: the random walk stage and a subsequent flooding/single-path routing phase. The first phase is a pure random walk or a directed walk, which meant to direct the message to a phantom source. And the other phase meant to deliver the message to the sink. When the source sends out a message, the message is unicasted in a random shift for a total of h_{walk} hops. After the h_{walk} hops, in phantom flooding the message is flooded using baseline (probabilistic) flooding. In phantom single-path routing, after the h_{walk} hops the message transmission turn into single-path routing. The ability of a phantom technique to enhance privacy is based on the ability of the random walk to place the phantom source (after h_{walk} hops) at a location far from the real source. The intention of the random walk is to send a message to a random location away from the real source. Nevertheless, if the network is more or less uniformly deployed, and to let those nodes randomly choose one of their neighbors with equal probability, then there is a large chance that the message path will loop around the source spot and branch to a random location not far from the source.

2. *Greedy random walk*

Y. Xi et al. proposed GROW (Greedy Random Walk), a two-way random walk, i.e., from both source and sink, to reduce the chance an eaves-dropper can collect the location information. They improve the delivery rate by using local broadcasting and greedy forwarding. The sink first sets up a path through random walk which serves as a receptor. Each packet from a source is then randomly forwarded until it reaches the receptor. At that point, the packet is forwarded to the sink through the pre-established path. A random walk greatly reduces the chance of packets being detected. Even if an eavesdropper happens to detect one packet, the next packet is unlikely to follow the same path, thus rendering the previous observation useless. In GROW, each time the sensor will pick up one of its neighbors which have not participated in the random walk. In this way, the random walk is always trying to cover an unvisited area using a greedy strategy. Moreover, it also eliminates local random walk and let both the source and sink initialize such a random walk to further improve the performance[95].

3. *Directed random walk*

J. Yao et al.[96] proposes a DROW (Directed Random Walk) method is to make it difficult for an adversary to backtrack hop-by-hop to the origin of the sensor communication. In DROW, the source sensor sends out a packet, the packet is unicasted to its parent node. When intermediate node receives a packet, it forwards to one of its parent nodes in a directed random fashion. DROW has several advantages compared to flooding-based phantom. DROW not only has smaller message latencies and lower energy costs, but also has better safety period when intermediate node has multi-parent node.

In addition, every sensor node can know the relative position of its neighbors by using DROW. Such knowledge can be obtained by following method. The value of level represents the number of hops that a node is from the base station along a particular path. A sensor node selects all neighbor nodes whose level value is less than its level value as its parent nodes. When a sensor node finds monitored object, it will report a message to the base station. The source sensor node sends out a packet, the packet is unicasted to its parent node. The intermediate node forwards the received packet to one of its parent nodes with equal probability. Each packet from source sensor node is forwarded until it reaches the base station in a directed random fashion.

8.6.3 Dummy massages strategies

To further protect the location of the data source, fake data packets can be introduced to perturb the traffic patterns that can be observed by the attacker.

1. *Cyclic entrapment method*

Reference [63] proposes a new cyclic entrapment method (CEM) that preserves the performance advantage of shortest path routing while also protecting the location of a source and adding a comparatively low cost in terms of additional message latency and energy. CEM generates some link loops in the network and misleads external attackers to these loops to protect the source location privacy. The CEM protocol is described in Fig. 8.11. Once a message is being routed along a path from the source to the base station and it encounters one of these pre-configured loops, the encountered loop will be activated and will begin cycling fake messages around the loop. When an attacker is trying to analyze the traffic and trace the message's path back to the source, it will need to select a direction to go on if it encounters a node that is a common node of both a loop and a correct path. Thus, it may make a wrong decision and be drawn into this loop. There is no way for an adversary to determine that whether the path they chose is true until they complete a cycle, thus the expected time for an adversary to find the correct path is increased. Therefore, it will take more time for the attackers to trace back to the source node by ensuring that a message's path is likely to cross multiple loops. Although CEM can obtain good safety period, the introduction of fake messages bring great energy waste. Moreover, the safety of CEM will be destroyed if the attacker has ability to observe traffic in a large area or to record nodes it has visited.

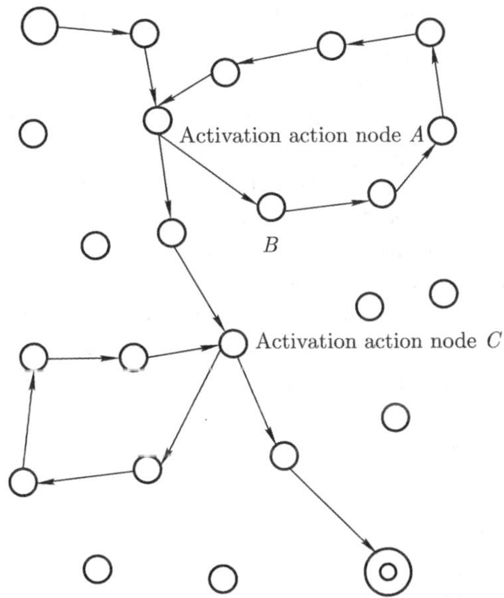

Fig. 8.11 Cyclic entrapment method.

2. *Source anonymity*

Reference [100] presents source anonymity for sensor networks under a global observer who may monitor and analyze the traffic over the whole network.

The basic idea of this approach is as follows. At first, network-wide dummy messages are employed. This is because it is unlikely to achieve source anonymity under such a strong attack model if all the traffic in the network is real event messages. Then, every node in the network sends out dummy messages with intervals following a certain kind of distribution. When a node detects a real event, it transmits the real event messages with intervals following the same distribution. By this means an attacker neither can identify the occurrence of a real event nor find out the location of the real event source. Moreover, two methods are introduced in order to reduce the extra overhead caused by dummy messages and guarantee the low real event report latency at the same time. Firstly, it relaxes the perfect source anonymity requirement and proposes a notion of statistically strong source anonymity for sensor networks. Secondly, project a realization scheme, called Fitted Probabilistic Rate scheme. Through selecting and controlling the probabilistic distribution of message transmission intervals, this scheme is able to makes the event notification delay is significantly reduced while keeping statistically strong source anonymity.

3. *Event source unobservability*

Reference [101] provides event source unobservability under a global attack model, where an attacker can hear and collect all the messages transmitted in the network at all the time. It promises that an attacker may neither discern the occurrence of a real event, nor find out the location of the real source. This is a stronger notion of privacy than traditional source location privacy that only hides the location of a real source.

The ideally result is to introduce carefully chosen dummy traffic to hide the real event sources and combine with mechanisms to drop dummy messages to prevent explosion of network traffic. To achieve the latter, they select some sensors as proxies that proactively filter dummy messages on their way to the base station. Since the problem of optimal proxy placement is NP-hard, it employs local search heuristics. To accurately locate proxies, two schemes are proposed: (a) Proxy-based Filtering Scheme (PFS) and (b) Tree-based Filtering Scheme (TFS). Simulation results show that these schemes not only quickly find nearly optimal proxy placement, but significantly reduce message overhead and improve message delivery ratio as well.

8.7 Conclusion

Wireless sensor networks have been proven lately a very useful type of networks. Although research on sensor networks security has achieved many notable results as addressed above, opportunities still remain in this area.

With the promotion of node's hardware performance and further research achievements, former accepted assumptions are more likely to be unsuited.

More challenges arise due to the continuous change of requirements. Areas are yet unexplored including optimization of security mechanisms in terms of resources and network environment, group re-keying infrastructure, and effective detection on DoS attacks.

Sensor network security is a critical issue but minimal research has been done compared to other aspects of WSNs. Sensor nodes are resource-constrained and embedded in physical environments, where unlimited resource for the calculation cannot be expected. A different technology from existing network security is required for WSNs.

As the development and research on this type of networks is still growing the need for including tools, such as trust or reputation is also growing. We believe these practices should be included in the design of a trust management system for WSN. According to the classification based on these best practices we have reviewed which existing approaches for trust or reputation systems for WSN take these practices into account. The success of the trust management system might depend on the adoption of the practices. By analyzing the existing approaches we have come to the conclusions that some of these practices are mostly overlooked by most of the proposals. This is the case, for example, of trust and reputation. In most of the cases they are considered jointly in order to build the trust or reputation systems. However, there are many other practices, such as trust of the base station, risk and importance and granularity, which are considered only by a few of the analyzed cases.

In WSNs, existing researches either only consider protecting source locations or only consider protecting sink location. It is necessary and challenging to design and implement strategies that can simultaneously protect location privacies of the source and the sink with low cost. In addition, how to protect location privacies of mobile base stations is also a challenging issue. It is obvious that a mobile base station can protect its location privacy well against external attackers; but it still needs to update its location information to the network, which may give more opportunities for the internal attackers to trace to it.

As future work, we intend to build lightweight trust management systems for WSN that include or at least consider as many of the best practices mentioned in this paper as possible. Besides, we will also analyze how the lack of a trust management system can affect the system. This will provide more accurate and reliable trust management systems for WSN. We also need to investigate the impact of source mobility, multiple sources, and base station mobility on location privacy protection issues. What's more, a real experiment is being designed to estimate the performance of algorithm. Moreover, applications based on node trust are being considered, such as routing, data aggregation and so on.

References

[1] Li P, Lin Y P, Zeng W N (2006) Search on security in sensor networks. Journal of Software, 17 (12): 2577 – 2588.

[2] Chong C Y, Kumar S P (2003) Sensor networks: Evolution, opportunities, and challenges. Proceeding of the IEEE, 91(8): 1247 – 1256.

[3] Huang L, Liu L (2008) Extended Watchdog Mechanism for Wireless Sensor Networks. Journal of Information and Computing Science, 3(1): 39 – 48.

[4] Roman R, Zhou J, Lopez J (2005) On the Security of Wireless Sensor Networks. In Proceedings of 2005 ICCSA Workshop on Internet Communications Security, LNCS 3482, pp. 681 – 690.

[5] Karlof C, Wagner D (2003) Secure routing in wireless sensor networks: Attacks and countermeasures. In Proceedings of the 1st IEEE International Workshop on Sensor Network Protocols and Applications Anchorage.

[6] Perrig A, Stankovic J, Wagner D (2004) Security in wireless sensor networks. Communications of the ACM, 47(6).

[7] Pathan ASK, Lee H-W, Hong C S (2006) Security in wireless sensor networks: issues and challenges. Advanced Communication Technology, 2006. ICACT 2006. The 8th International Conference, 2(20 – 22).

[8] Nagai C H (2006) Intrusion Detection for Wireless Sensor Networks. Ph.D. Term 2 Paper, The Chinese University of Hong Kong. Department of Computer Science and Engineering. www.cse.cuhk.edu.hk/~lyu/student/phd/edith/edith_term2.pdf. Accessed 19 June, 2011.

[9] Islam M S, Khan R H, Bappy D M (2010) A Hierarchical Intrusion Detection System in Wireless Sensor Networks. IJCSNS International Journal of Computer Science and Network Security, 10(8).

[10] Wood A D, Stankovic J A (2002) Denial of Service in Sensor Networks. Computer, 35(10): 54 – 62.

[11] Anderson R, Kuhn M (1996) Tamper Resistance a Cautionary Note. In Proceedings of 2nd Usenix Workshop Electronic Commerce, pp. 1 – 11.

[12] Johnson D B, Maltz D A (1996) Dynamic Source Routing in Ad Hoc Wireless Networks. Mobile Computing, vol. 353, T. Imielinski and H. Korth, eds., Kluwer Academic, pp. 153 – 181.

[13] Karp B, Kung H T (2000) GPSR: Greedy Perimeter Stateless Routing for Wireless Networks. In Proc. of 6th Ann. Int'l Conf. Mobile Computing and Networking (MobiCom 2000), ACM Press, New York, pp. 243 – 254.

[14] Perkins C E, Bhagwat P (1994) Highly Dynamic Destination-Sequenced Distance-Vector Routing (DSDV) for Mobile Computers. In Proc. of SIGCOMM, ACM Press, New York, pp. 234 – 244.

[15] Cheung S, Levitt K N (1997) Protecting Routing Infrastructures from Denial of Service Using Cooperative Intrusion Detection. Proc. Workshop New Security Paradigms, ACM Press, New York, pp. 94 – 106.

[16] Schuba C L (1997) Analysis of a Denial of Service Attack on TCP. In Proc. of IEEE Symp. Security and Privacy, IEEE Press, Piscataway, N. J., pp. 208 – 223.

[17] Perrig A, Szewczyk R, Wen V, Culler D, Tygar J D (2001) SPINS: Security protocols for sensor networks. In Proc. of the 7th Annual Int'l Conf. on Mobile Computing and Networks. ACM Press, New York, pp. 189 – 199.

[18] Zhu S, Setia S, Jajodia S (2003) LEAP: Efficient security mechanisms for large-scale distributed sensor networks. In Proc. of the 10th ACM Conf. on Computer and Communications Security (CCS 2003), pp. 62 – 72.

[19] Chan H, Perrig A, Song D (2003). Random key predistribution schemes for sensor networks. In Proc. of the IEEE Symp. on Research in Security and Privacy. IEEE Computer Society, pp. 197 – 213.

[20] Li P, Lin Y P (2006) Search on security in sensor networks. Journal of Software, 17(12): 2577 – 2588.

[21] Zhang X Y, Heys H M, Li C (2010) Energy Efficiency of Symmetric Key Cryptographic Algorithms in Wireless Sensor Networks. Communications (QBSC), 2010 25th Biennial Symposium on Digital Object Identifier: 10.1109/BSC.2010.5472979

[22] Goldwasser S, Micali S (1984) Probabilistic encryption. Journal of Computer Security, 28: 270 – 299.

[23] Ren K, Yu S, Lou W, Zhang Y (2009) Multi-user broadcast authentication in wireless sensor networks. IEEE Transactions on Vehicular Technology, 58(8): 4554 – 4564.

[24] Menezes A J, Van Oorschot P C., Vanstone S A (1996) Handbook of applied cryptology. Retrieved from http:/www.cacr.math.uwaterloo.ca/hac/about/chap10.pdf. Accessed 19 June, 2011.

[25] Shih W, Hu W, Corke P, Overs L (2008) A public key technology platform for wireless sensor networks. In Proceedings of the 6th ACM conference on Embedded Network Sensor Systems, pp. 447–448.

[26] Schmidl G, Rossi F (2010) A-Code: A New Crypto Primitive for Securing Wireless Sensor Networks. CNSA 2010, CCIS 89, pp. 452–462. Springer-Verlag, Heidelberg.

[27] Huang Q, Cukier J, Kobayashi H, Liu B D, Zhang J Y (2003) Fast authenticated key establishment protocols for wireless sensor networks. In Proc. of the 2nd ACM Int'l Conf. on Wireless Sensor Networks and Applications. San Diego: ACM press, New York, pp. 141 – 150.

[28] The official website of the ZigBee alliances. Retrieved from http://www.zigbee.org. Accessed 19 June, 2011.

[29] Szczechowiak P, Oliveira L, Scott M, Collier M, Dahab R (2008) Nanoecc: testing the limits of elliptic curve cryptography in sensor networks. In Proceedings of European Conference on Wireless Sensor Networks (EWSN '08), pp. 305–320, Springer, New York.

[30] Azarderskhsh R, Reyhani-Masoleh Arash (2011) Secure Clustering and Symmetric Key Establishment in Heterogeneous Wireless Sensor Networks. EURASIP Journal on Wireless Communications and Networking. Volume 2011, Article ID 893592, 12 pages. DOI: 10.1155/2011/893592.

[31] Malan D J, Welsh M, Smith M D (2004) A public-key infrastructure for key distribution in Ting OS based on elliptic curve cryptography. Retrieved from http://airclic.eecs.har vard.edu/publications/secon04. pdf. Accessed 19 June,2011.

[32] Crossbow I (2008). Technology MICA2: Wireless measurement system. http://www.xbow.com/Products/Product_pdf_files/Wireless_pdf/6020-0042-04_MICA2.pdf. Accessed 19 June, 2011

[33] Du W L, Wang R H, Ning P (2005) An efficient scheme for authenticating public keys in sensor networks. In Proc. of the 6th ACM Int'l Symp. on Mobile Ad Hoc Networking and Computing (MobiHoc 2005), pp. 58 – 67. ACM press, New York.

[34] Su Z, Lin C, Feng F J, Ren F Y (2007) Key management schemes and protocols for wireless sensor networks. Journal of Software, 18(5): 1218-1231.

[35] Eschenauerl L, Gligor V D (2002) A key-management scheme for distributed sensor networks. In proceedings of the 9th ACM Conference on Computer and Communications Security (CCS'02).

[36] Du W L, Deng J, Han Y H, Chen S G, Varshney P K (2004) A Key Management Scheme for Wireless Sensor Networks Using Deployment Knowledge. In Proceedings of IEEE INFOCOM 2004.

[37] Blundo C, Santis A D, Herzberg A, Kutten S, Vaccaro U, Yung M (1992) Perfectly Secure key distribution for dynamic conferences. In Crypto.

[38] Du W L, Ning P (2003) Establishing pairwise keys in distributed sensor networks. CCS'03.

[39] Du W L, Deng J, Han Y S, Varshney P, Katz J, Khalili A (2003) A Pairwise Key Pre-Distribution Scheme for Wireless Sensor Networks. CCS'03.

[40] Curiac D, Plastoi M (2009) Combined Malicious Node Discovery and Self-Destruction Technique for Wireless Sensor Networks. IEEE DOI 10.1109/SENSORCOMM.2009.72.

[41] Dong H, Guo Y, Yu Z Q, Chen H (2009) A Wireless Sensor Networks Based on Multiangle Trust of Node. 2009 International Forum on Information Technology and Applications.

[42] Viljanen L (2005) Towards an Ontology of Trust. In Proceedings of the Trust Bus 2005, LNCS 3592, pp. 175 – 184.

[43] Crosby G V, Pissinou N, Gadze J (2006) A Framework for Trust-based Cluster Head Election in Wireless Sensor Networks. In Proceedings of Second IEEE Workshop on Dependability and Security in Sensor Networks and Systems, pp. 13 – 22.

[44] Ganeriwal S, Balzano L K, Srivastava M B (2008) Reputation-Based Framework for High Integrity Sensor Networks. ACM Trans. Sens. Netw. 4, 1 – 37.

[45] Tang W, Hu J B, Chen Z (2005) Research on a Fuzzy Logic-Based Subjective Trust Management Model. J. Comput. Res. Develop. 42, 1654 – 1659.

[46] Krasniewski M, Varadharajan P, Rabeler B, Bagchi S, Hu Y C (2005) TIBFIT: Trust Index Based Fault Tolerance for Arbitrary Data Faults in Sensor Networks. In Proceedings of the 2005 International Conference on Dependable Systems and Networks, pp. 672 – 681.

[47] Song F, Zhao B H (2008) Trust-Based LEACH Protocol for Wireless Sensor Networks. In Proceedings of the Second International Conference on Future Generation Communication and Networking, pp. 202 – 207.

[48] Tanachaiwiwat S, Dave P, Bhindwale R, Helmy A (2003) Secure Locations: Routing on Trust and Isolating Compromised Sensors in Location-aware Sensor Networks. In Proceedings of the SenSys, pp. 324 – 325. Sensors 2011, 111360.

[49] Tanachaiwiwat S, Dave P, Bhindwale R, Helmy A (2004) Location-Centric Isolation of Misbehavior and Trust Routing in Energy-Constrained Sensor Networks. In Proceedings of the 23rd IEEE International Performance, Computing, and Communications Conference, pp. 463 – 469.

[50] Hur J, Lee Y, Hong S M, Yoon H (2005) Trust Management for Resilient Wireless Sensor Networks. In Proceedings of the 8th International Conference on Information Security and Cryptology, pp. 56 – 68.

[51] Hur J, Lee Y, Yoon H, Choi D, Jun S (2005) Trust Evaluation Model for Wireless Sensor Networks. In Proceedings of the 7th International Conference on Advanced Communication Technology, pp. 491 – 496.

[52] Almenarez F, Marin A, Diaz D, Sanchez J (2006) Developing a Model for Trust Management in Pervasive Devices. In Proceedings of 4th IEEE Annual International Conference on Pervasive Computing and Communications, pp. 267 – 271.

[53] Almenarez F, Marin A, Campo C, Garcia R C (2004) PTM: A Pervasive Trust Management Model for Dynamic Open Environments. In Proceedings of the 1st Workshop on Pervasive Security, Privacy and Trust.

[54] Almenarez F, Marin A, Campo C, Garcia RC (2005) Trust AC: Trust-based Access Control for Pervasive Devices. In Proceedings of the 2nd International Conference on Security in Pervasive Computing, pp. 225 – 238.

[55] Hsieh M Y, Huang Y M, Chao H C (2007) Adaptive Security Design with Malicious Node Detection in Cluster-Based Sensor Networks. Comput. Commun. 30, pp. 2385 – 2400.

[56] Marmol F G, Perez G M (2010) Towards Pre-standardization of Trust and Reputation Models for Distributed and Heterogeneous Systems. Comput. Stand. Interfaces 2010, 32, pp. 185 – 196.

[57] Lopez J, Roman R, Agudo I, Fernandez C G (2010) Trust Management Systems for Wireless Sensor Networks: Best Practices. Comput. Commun. 33, pp. 1086 – 1093.

[58] Li J L, Gu L Z, Yang Y X (2009) A New Trust Management Model for P2P Networks. J. Beijing Univ. Posts Telecommun. 32, pp. 71 – 74.

[59] Li J L, Gu L Z, Yang Y X (2009) A New Trust Management Model for P2P Networks with Time Self-Decay and Subjective Expect. J. Electron. Inf. Technol. 31, pp. 2786 – 2790.

[60] Li L, Fan L, Hui H (2009) Behavior-Driven Role-Based Trust Management. J. Softw. 20, pp. 2298 – 2306.

[61] Feng R, Xu X F, Zhou X, Wan J W (2011) A Trust Evaluation Algorithm for Wireless Sensor Networks Based on Node Behaviors and D-S Evidence Theory. Sensors 2011, 11, pp. 1345 – 1360; doi:10.3390/s110201345.

[62] Carman D, Kruus P, Matt B (2000) Constraints and approaches for distributed sensor network security. Technical Report # 00-010. NAI Labs. http://www.csee.umbc.edu/courses/graduate/CMSC691A/Spring04/papers/nailabs_report_00-010_final.pdf. Accessed 9 December, 2010.

[63] Yi O, Le Z Y, Chen G, Ford J, Makedon F (2006) Entrapping adversaries for source protection in sensor networks. World of Wireless, Mobile and Multimedia Networks, WoWMoM 2006. International Symposium.

[64] Doumit S, Agrawal D P (2003) Self-Organized criticality and stochastic learning-based intrusion detection system for wireless sensor networks. MILCOM 2003—IEEE Military Communications Conf., 22(1): 609 – 614.

[65] Marti S, Giuli T, Lai K, Baker M (2000) Mitigating routing misbehavior in mobile ad hoc networks. In proc. ACM MobiCom, pp. 255-265.

[66] Khalil I, Bagchi S, Shroff N B (2005) LiteWorp: a lightweight countermeasure for the wormhole attack in multihop wireless networks. In International Conference on Dependable Systems and Networks (DSN), pp. 612 – 621.

[67] Wang X, Xu J, Wang J (2009) Detection and location of malicious nodes based on source coding and multi-path transmission in WSN. High Performance Computing and Communications, HPCC '09. 11th IEEE International Conference.

[68] Wang X, Wong J (2007) An End-to-end Detection of Wormhole Attack in Wireless Ad Hoc Networks. In Proceedings of the 31st Annual International Computer Software and Applications Conference-Vol. 1-(COMPSAC 2007).

[69] R. da Silva A, Martins M, Rocha B (2005) Decentralized Intrusion Detection in Wireless Sensor Networks. In Proceedings of the 1st ACM international workshop on Quality of service & security in wireless and mobile networks (Q2SWinet' 05), pp. 16 – 22.

[70] Tseng C-Y, Balasubramanyam P, Ko C, Limprasittiporn R, Rowe J, Levitt K (2003) A specification-based intrusion detection system for AODV. In Proceedings of the 1st ACM workshop on Security of Ad Hoc and sensor networks.

[71] Momani M, Agbinya J, Navarrete G P, Akache M (2006) A New Algorithm of Trust Formation in Wireless Sensor Networks. In The 1st IEEE International Conference on Wireless Broadband and Ultra Wideband Communications (AusWireless '06).

[72] Yao Z Y, Kim D Y, Lee I (2005) A security framework with trust management for sensor networks. In Proc. of the 1st IEEE/CREATE-NET Workshop on Security and QoS in Communication Networks Athens. Piscataway, IEEE Computer Society, pp. 190 – 198.

[73] Yao Z Y, Kim D, Doh Y (2006) PLUS: Parameterized and localized trust management scheme for sensor networks security. In Proc. of the IEEE Int'l Conf. on Mobile Ad Hoc and Sensor Systems (MASS). Piscataway, IEEE Computer Society, pp. 437 – 446.

[74] Shaikh R A, Jameel H, Lee S, Rajput S, Song Y J (2006) Trust management problem in distributed wireless sensor networks. In Proc. of the RTCSA. Piscataway, IEEE Computer Society, pp. 411-414.

[75] Ryutov T, Neuman C (2007) Trust based approach for improving data reliability in industrial sensor networks. In Etalle S, eds. In Proc. of the IFIP Int'l Federation for Information, 238: 349 – 365.

[76] Probst M J, Kasera S K (2007) Statistical trust establishment in wireless sensor networks. In Proc. of the Int'l Conf. on Parallel and Distributed Systems. IEEE Computer Society, pp. 1–8.

[77] Agah A, Das S K, Basu K (2004) A game theory based approach for security in wireless sensor networks. In Proc. of the IEEE Int'l Conf. on Performance, Computing and Communications. Piscataway, IEEE Computer Society, pp. 259–263.

[78] Blaze M, Feigenbaum J, Lacy J (2002) Decentralized Trust Management. In Proceedings of the 1996 IEEE Symposium on Security and Privacy, pp. 164–173.

[79] Ellison C M, Frantz B, Lampson B, Rivest R, Thomas B M, Ylonen T (2010) Simple Public Key Infrastructure Certificate Theory. Retrieved from http://www. ietf.org/ietf/1id-abstracts.txt. Accessed 9 December, 2010.

[80] Li N H, Mitchell J C (2003) RT: A Role-Based Trust-management Framework. In Proceedings of the 3rd DARPA Information Survivability Conference and Exposition, pp. 201–212.

[81] Li N H, Mitchell J C, Winsborough W H (2005) Beyond Proof-of-Compliance: Security Analysis in Trust Management. J. ACM, 52, 474–514.

[82] Josang A, Ismail R (2002) The Beta Reputation System. In proceeding of 15th Bled Electronic Commerce Conference, pp. 17–19.

[83] Srinivasan A, Teitelbaum J, Wu J (2006) Distributed Reputation-based Beacon Trust System. In Proceedings of 2nd IEEE International Symposium on Dependable, Autonomic and Secure Computing, pp. 277–283.

[84] Josang A, Ismail R, Boyd C (2007) A Survey of Trust and Reputation Systems for Online Service Provision. Decision Support Systems, pp. 618–644.

[85] Wagner D (2004) Resilient aggregation in Sensor Networks, In Proceedings of the 2nd ACM workshop on Security of Ad hoc and Sensor Networks, pp. 78–87.

[86] Capkun S, Srivastava M, Cagalj M (2006) Securing localization with hidden and mobile base stations. In INFOCOM 2006.

[87] Lazos L, Poovendran R (2004) SeRLoc: Secure range-independent localization for wireless sensor networks. In ACM WiSe 2004.

[88] Lazos L, Poovendran R, Capkun S (2005) ROPE: Robust position estimation in wireless sensor networks. In IPSN 2005.

[89] Li Z, Trappe W, Zhang Y, Nath B (2005) Robust statistical methods for securing wireless localization in sensor networks. In IPSN 2005.

[90] Liu D, Ning P, Du W (2005) Attack-resistant location estimation in sensor networks. In IPSN 2005.

[91] Sastry N, Shankar U, Wagner D (2003) Secure verification of location claims. In ACM Wise 2003.

[92] Capkun S, Hubaux J P (2005) Secure positioning of wireless devices with application to sensor networks. In INFOCOM 2005.

[93] Vora A, Nesterenko M (2004) Secure location verification using radio broadcast. In International Conference on Principles of Distributed Systems.

[94] Kamat P, Zhang Y, Trappe W (2005) Enhancing Source-Location Privacy in Sensor Network Routing. In Proc. of IEEE ICDCS' 05.

[95] Xi Y, Schwiebert L, Shi W (2006) Preserving location privacy in monitoring based wireless sensor networks. In Proceedings of the 2th International Workshop on Security in Systems and Networks (SSN'06). IEEE Computer Society.

[96] Yao J, Wen G (2008) Preserving Source-Location Privacy in Energy Constrained Wireless Sensor Networks. In Proc. of ICDCS 2008 workshops. IEEE Press.

[97] Ozturk C, Zhang Y, Frappe W (2004) Source-location privacy for networks of energy constrained sensors. In Proceedings of 2nd IEEE Workshop off Software Technologies for Future Embedded and Ubiquitous Systems (WSTFEUS04), pp. 68 – 72.

[98] Braginsky D, Estrin D (2002) Rumor routing algorthim for sensor networks. In Proceedings of the 1st ACM international workshop on wireless sensor networks and applications.

[99] Eugster P T H, Guerraoui R, Handurukande S B, Kouznetsov P, Kermarrec A-M (2003) Lightweight probabilistic broadcast. ACM Transactions on Computer Systems (TOCS), 21(4): 341 – 374.

[100] Shao M, Yang Y, Zhu S (2008) Towards Statistically Strong Source Anonymity for Sensor Networks. In Proc. of IEEE INFOCOM' 08.

[101] Yang Y, Shao M, Zhu S (2008) Towards event source unobservability with minimum network traffic in sensor networks. In Proc. of ACM WiSec'08.

Chapter 9
Security in Wireless Mesh Networks

Chung-wei Lee[1]

Abstract

The rapid emergence of wireless electronics such as iPhone and iPad has changed the way people communicate with each other. While wireless service providers continue to expand the capacity of their network infrastructures, one of the key components — wireless mesh network (WMN) — is expected to dominate the wireless interconnection and access networks. With wireless services integrated into our work and home activities, WMN security issues become evident. This chapter provides an extensive coverage on the security challenges, requirements, attacks, and countermeasure mechanisms that are related to wireless mesh networks. The treatment of these subjects is based on both theoretical analysis and practical application. In addition, cutting-edge wireless mesh network research projects and commercial products are discussed to provide technical insights for researchers and practitioners.

9.1 Introduction

Since the early work of radio transmission in the late 1800s, wireless communication technology has been advanced dramatically. While the early progress was mainly contributed by ingenious scholars, researchers, and inventors (such as Hertz, Maxwell, Edison, etc.), recent rapid development seemed to be triggered by the demand of ever-growing wireless consumers. In the past 20 years, wireless computers and cellular phones have evolved from being luxury products to daily necessities. They are not only more powerful in capacity but also much smaller in size. With this trend, mobile devices equipped with high-throughput wireless communication capability will soon dominate our consumption on networking resources.

1 Department of Computer Science, University of Illinois at Springfield, One University Plaze, MS UHB 3100, Springfield, Illinois 62703-5407, USA.

Wireless mesh networks are expected to provide interconnected network services to end users who rely mainly on wireless connections for everyday communication needs. These may include smartphone voice conversation, web browsing, data transactions, or cloud clients' extensive access to cloud services. Although wireless services have integrated into our work and home activities, the need for strong wireless security becomes evident. Because wireless mesh networks are a major part of this network service, its security features deserve to be studied and understood in detail.

A typical wireless mesh network (WMN) is a collection of wireless local area networks (WLANs) that are interconnected together to form a meshed WLAN network[1]. The dominant WLAN technology, at the time of this writing, is the IEEE 802.11 WLAN (a.k.a. Wi-Fi). IEEE 802.11 is a WLAN standard series. That consists of many versions/amendments, with 802.11b/802.11g/802.11n the most well-known ones. While a single WLAN can provide certain wireless services to users in a close proximity, the federation of nearby WLANs can enhance the locally constrained service to smooth global roaming and reaching the abundant Internet/Web services. Interconnecting a group of WLANs was traditionally done by wired networks. However, this approach is costly and inflexible[2]. In contrast, WMNs are able to replace the wired interconnection networks with a wireless version, and thus offer an alternative solution which is inexpensive and is easily adapted to surroundings. Key functions of a WMN include automatic topology discovery, dynamic routing, quality of service, and security.

Many industries and organizations have already adopted (or have a strong interest to adopt) WMNs for their communication infrastructure need. For example, most health care professionals (such as doctors and nurses) need to move frequently from place to place, while at the same time have access to patient information in a secure and reliable manner. WMNs can be deployed in hospitals and health care facilities to serve as the communication backbone. WMNs are designed to be flexible in deployment, robust, and secure in wireless connections for such an environment. Another popular community for WMN employment is in educational institutions. Many colleges and universities were actually the early adopters of WMN technology because of their involvement in creating and developing experimental WMN systems. Another high-profile WMN example is the One Laptop Per Child (OLPC) project[3]. The major goal of OLPC is to allow disadvantaged school children in developing (or the least developed) countries to have economical laptop computers and access to the resourceful Internet. The key role that WMNs play in this project is to connect these wireless laptop computers via ad hoc manner so that students can collaborate with each other and share the Internet access connections. While the social or political success of OLPC is still debatable, the technical importance of WMNs in the project is undeniable.

From the point of view of WMN security, a significant real-world application is for the public safety and disaster recovery (PSDR) wireless communication system[4]. During emergency or disaster situations, most wire-based communication infrastructures would be out of function and the repair of them would not be able to be completed in time. The fast deployment of WMNs requires no wire layout and is friendly to terrain. Yet, they offer reliable mobile communication through wireless channels that is capable of providing confidential connection, message integrity, and strong authentication service.

9.2 Wireless Mesh Networks (WMN) Characteristics

The major components in a wireless mesh network are generally classified as mesh client, mesh router, and gateway. However, IEEE 802.11s[5] uses different names[6] (in addition to the standard IEEE 802.11 station): mesh station (Mesh STA)/point (MP), mesh access point (MAP), and mesh portal point (MPP). Table 9.1 shows the mapping between the general terms and the terms specified in IEEE 802.11s and their respective description.

Table 9.1 Wireless Mesh Network Components

General Terms	IEEE 802.11s Terms	Description
Mesh Client	Station (STA)	An end-user IEEE 802.11 wireless device (such as a computer or smartphone) that makes service requests
Mesh Router	Mesh Station (Mesh STA) or Mesh Point (MP)	A wireless device that participates in building wireless mesh networks, forwarding frames to peers, and supporting other relevant mesh service functions such as security and management
	Mesh Access Point (MAP)	A mesh station that supports IEEE 802.11 access point (AP) functions which provide network access to non-mesh stations wirelessly
Gateway	Mesh Portal Point (MPP)	A mesh station that interconnects with non-802.11 networks and provides access to external networks (e.g., Internet)

Depending on the application situations, the deployment configuration of a wireless mesh network can be very flexible. For example, in a typical airport building consisting of a check-in area and two terminals as shown in Fig. 9.1, multiple MAPs are deployed to provide network access to airline staff members as well as customers. To support efficient and robust wireless routing in this WMN, several MPs (or mesh STAs) are positioned in strategic locations for performance maximization. For all network traffic to/from the external networks, two MPPs having high-speed external connections (wired or wireless) are utilized.

As wireless mesh networks are interconnected wireless local networks that can provide Internet data access as well as real-time voice and video services, they have a strong relationship with the rest of the wired and wireless network world. With the intertwined functions and connections, it is imperative to understand their differences.

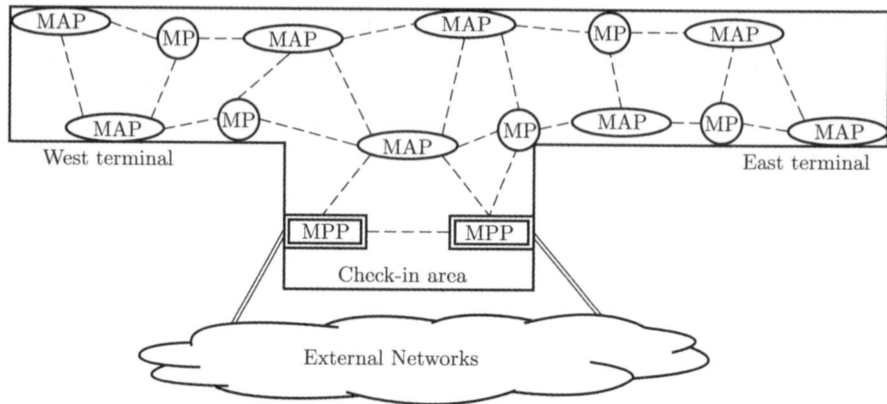

Fig. 9.1 WMN configuration example.

9.2.1 WMN vs. Cellular Networks

The three major differences between a WMN and a cellular network are frequency spectrum, network topology/configuration, and routing process[7]. In a cellular phone network system, consumers pay for their air time because their wireless communication is conducted through licensed radio frequency bands. Service vendors pay a license fee to government agencies for the privilege of utilizing particular radio bands. On the other hand, Wi-Fi-based WMNs utilize unlicensed frequency bands and therefore do not pay for the spectrum usage. In general, the radio transmission range for a cellular device is longer than the one in a WMN.

Regarding the network topology and configuration, a typical wireless cellular network system divides a large geographic region into many small areas where each such small area is called a "cell". A cell is centered on a base station (a.k.a. antenna tower) whose main task is to provide two-way communication service to users who are in close proximity. The interconnection among cells (may have multiple layers) are wired networks. Therefore, a cellular user's connection to the outside network (including Internet) is only one-hop (or last mile) wireless communication. On a WMN, a mesh point (MP) is first wirelessly connected to a nearby mesh access point (MAP), then one or more hops away a mesh portal point (MPP) can be reached and finally gain access to the outside network. That is, multi-hop wireless connections are normal in WMNs.

In a one-hop cellular network, there are no routing concerns. In WMNs, with multi-hop topology, selecting the best routing path from one point to another becomes an important issue in terms of performance and security. The wireless radio transmission channels are inherently insecure and vulnerable to attacks. Therefore, the design of WMN must address not only the performance issue but also the security concern. In many situations, security considerations can compromise performance and vice versa. For example, if strong security features are desired, then more computational resources (time and energy) are required which make the system run slow and easily exhaust the battery power.

9.2.2 WMN vs. Internet

Traditionally the Internet is considered a network of networks which consists of mostly wired networks. On a WMN, all communication channels are supported by wireless radio transmission. Besides the easy eavesdropping in wireless channels, the neighboring nodes can change dynamically in WMNs. That is, Internet routers are less vulnerable than WMN access points in terms of identity confirmation. Without the physical protection (which is enjoyed by most Internet routers in a locked closet or building), WMN access points need stronger mutual authentication (ID confirmation) before they accept new neighbor nodes. If an adversary successfully joins a WMN (i.e., proper authentication failed), it can launch more damaging passive and active attacks in the future which may threaten the existence of a good-faith federation of WMNs and compromise secret messages exchanged inside the network.

9.2.3 WMN vs. Mobile Ad Hoc Networks

While both Wi-Fi-based WMN and mobile Ad Hoc network (MANET) use unlicensed frequency bands for wireless communication and carry out multi-hop routing for end-to-end message exchange, there are other important differences between them.

- Most WMN mesh access points and mesh portal points are equipped with high performance antennas with multiple input multiple output (MIMO) capability. This enables them to utilize multiple frequency bands simultaneously and achieve relatively higher bandwidth (transmission speed) than typical MANET nodes.
- Most WMN mesh access points and mesh portal points are deployed strategically so that they have constant power supply to sustain their high throughput wireless backbone communication tasks. That is, they are expected to have very low or no mobility. In MANETs, all wireless devices are considered to have potentially high mobility which can cause performance degradation or service interruption.
- The key feature of a MANET is that it is formed without a fixed infrastructure. With the expected low mobility, WMN mesh access points can form stable wireless infrastructures to facilitate high throughput.

- In a MANET, most traffic flows are among user/node themselves. But a significant amount of traffic flow in WMNs is expected to go through the mesh access portal to exchange messages with users and servers in outside networks.

9.3 WMN Security Vulnerabilities

Wireless mesh networks are inherently vulnerable to threats and attacks known to wireless communication and mobile ad hoc networks. Some of them can be prevented or protected by applying the right choice of security mechanisms and services, others are difficult to defend by their nature. A good example is radio jamming at the physical layer. If an adversary has sufficient jamming equipments and power supply, the available wireless channels can be rendered useless. There are basic anti-jamming techniques such as spread spectrum (including frequency hopping, direct sequence, orthogonal frequency division multiple access), that can only make the jamming effort difficult but not fundamentally eliminate them, especially when adversaries possess powerful and long-lasting resources. In this section, the vulnerabilities that are most relevant to wireless mesh networks will be discussed.

1. *Compromised Mesh Stations*

While fast deployment is a positive feature of using WMNs, it comes with the risk of lacking physical protection of deployed mesh stations[8]. For example, in the case of military deployment in a battlefield, each soldier can be equipped with a WMN-capable wireless device for communication and control. They are thus considered as mesh points in the WMNs. Most mesh access points are expected to be installed on vehicles (air or ground) with certain physical protection, soldier-based mesh points are exposed to the danger of being captured or destroyed. When these unfortunate events occur, one of the following security threats become imminent[7].

- A mesh point suddenly disappears from the WMN.
- The adversary retrieves secrets, keys, routing information, in-transition packets.
- The adversary modifies the mesh point's routing parameters.
- The mesh point is cloned (duplicated) for future attacks.

The consequences of some of these cases can be catastrophic and disrupt the military action.

2. *Routing Threats and Attacks*

Since wireless mesh networks use multi-hop routing for end-to-end data packets delivery, the routing process must be made robust and secure so that the network can operate in a satisfactory manner. At the core of this, routing information is expected to be exchanged with adequate and efficient protection.

The following key principles need to be addressed[9].

- When routing messages are received, they should be validated by proper authentication procedure to make sure that they come from legitimate mesh nodes (i.e., these routing messages are not fabricated by adversaries).
- Received routing messages should be checked for message integrity to make sure that they have not been altered during transmission (i.e., attackers cannot modify the messages without being detected).

3. *Denial of Service*

Denial-of-Service (DoS) is the form of attacks that target on resource availability. This kind of attack may occur at different layers and/or areas of network structure, and is one of the most difficult attacks to thwart. For example, a DoS attack may take place at the physical layer in the form of radio channel jamming. Or, if an adversary takes control of a previously legitimate mesh point, it can launch DoS at the medium access control (MAC) layer by continuously requesting transmission privilege (implicitly or explicitly) thus hogging bandwidth resource to prevent others from transmitting.

4. *Wormholes, Gray Holes, Black Holes*

Wormhole, gray hole, and black hole attacks intend to re-direct network traffic to the advantage of adversary nodes. They are executed by providing false routing information to the rest of network nodes so that packets/traffic can be attracted to the adversary nodes. For example, in Fig. 9.2, node A represents a black hole. All arriving packets to node A will be discarded without notice. This misbehavior can disrupt normal network operation which causes significant waste of resources. An example of a gray hole is shown in node B where it selectively discards packets so that legitimate nodes can hardly detect its existence. However, damage can still be done because of short-term and/or long-term network performance degradation. Among these three types of attacks, the wormhole attack is probably the most harmful one because it has the potential to threaten routing integrity and conduct traffic analysis for secret revealing and launch large scale DoS attacks at a later time. Nodes C and D in the figure demonstrate this type of attack.

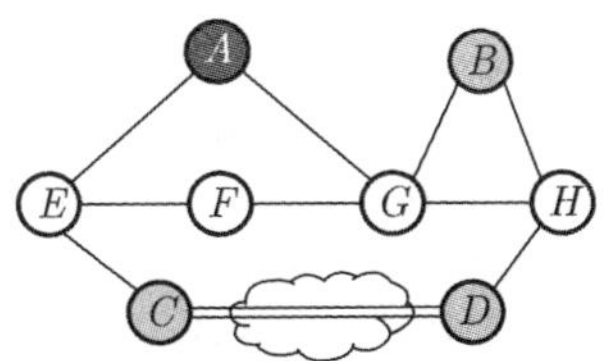

Fig. 9.2 Wormholes, gray holes, and black holes.

9.4 WMN Defense Mechanisms

However there are many general wireless network defense mechanisms and systems, not all of them are suitable for wireless mesh networks. In this section, the most relevant ones are discussed. They include the IEEE 802.11i security model, advanced authentication and key management, and sophisticated path selection and routing schemes.

9.4.1 IEEE 802.11i Security Model

In IEEE 802.11i[10] based wireless networks such as those complied with Wi-Fi Alliance's Wi-Fi Protected Access II (WPA2) [11], the access control and authentication process are implemented by the integration of three protocols: IEEE 802.1X, EAP, and RADIUS [12]. In addition, a robust security network (RSN) can be created by proper associations using a four-way handshake procedure. Strong data confidentiality and integrity are provided by Advanced Encryption Standard Counter Mode — CBC MAC Protocol (AES-CCMP).

A typical IEEE 802.11i operation includes four phases [13,14] as shown in Fig. 9.3.

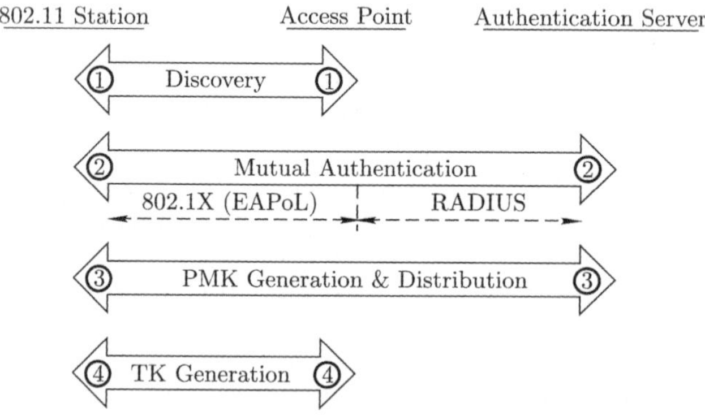

Fig. 9.3 IEEE 802.11i operation phases.

In the phase of discovery, an 802.11 station (STA) and an access point (AP) exchange messages that facilitate the negotiation of security features. At the end of this phase, they agree upon the authentication and encryption algorithms that can be used for the remaining of the process. In the second phase, STA and the authentication server (AS) exchange EAP messages. These messages are encapsulated in either EAPoL (EAP over LAN, between STA and AP) or RADIUS protocol (between AP and AS). If both STA and AS successfully authenticate each other, a Master Key (MK) will be generated on both sides and serve as a shared secret. In the third phase, the Pairwise Master Key (PMK) is generated in STA and AS. While the STA keeps the

PMK to itself, AS needs to deliver PMK to the AP so that STA and AP can have a shared key (i.e., PMK). While PMK is now possessed by both STA and AP, it is not used directly for the link-level communication. Instead, temporal keys (TK) are derived from PMK and used for further message encryption and integrity check.

1. *IEEE 802.1X*

IEEE 802.1X[15] is a port-based access control to protect network connections in non-secure environments. It divides all network components into three different roles: supplicant, authenticator, and authentication server. In a wireless network setting, supplicants are mobile devices (e.g., computers, smartphones, ...) that wish to access the network resources and therefore have to be authenticated before they are allowed in. Authenticators in Wi-Fi typically reside in access points or wireless routers which are the contact points for infrastructure-based Wi-Fi networks. Authentication servers are the decision making security entities in the authentication process. They usually hold the identifiers and credentials of legitimate clients (supplicants), and grant or deny access requests made by supplicants. While the logical authentication message exchange occurs between the supplicants and authentication servers, the real communication path consists of two segments: "supplicant $\leftrightarrow$ authenticator" and "authenticator $\leftrightarrow$ authentication server". That is, authenticators act as message relays and provide an extra layer of security defense. The existence of authenticators also allows efficient management of the authentication server systems.

2. *Extensible Authentication Protocol* (*EAP*)

EAP [16] is a framework that provides common authentication functions and negotiation. It consists of four main types of messages: request, response, success, and failure. The request and response messages are used to carry authentication-specific information, and the success and failure messages are indications of authentication results. The Internet Engineering Task Force (IETF) has published numerous Request for Comments (RFCs) and Internet Drafts that extend the basic EAP to other security protocols. For example, EAP is frequently used with upper-layer authentication protocols such as transport layer security (TLS/SSL) and Kerberos authentication system. EAP-TLS [17] is chosen as the de facto 802.11i authentication scheme because of the extensive deployment base of TLS in the current Internet. It defines the TLS handshaking procedure over EAP, thus adapts to the security models of 802.1X and 802.11i. A successful EAP-TLS message exchange accomplishes both mutual authentication and key derivation between STA and AS. The mutual authentication is achieved by exchanging and verifying digital certificates (a public key technique) on both sides. Then, the Master Key (MK) and Pairwise Master Key (PMK) can be derived and put into use for creating more link-level temporal keys. Another well-known EAP extension is the EAP-SIM [18] protocol. It provides the mechanism to incorporate EAP in the

Subscriber Identity Module (SIM) in Global System for Mobile Communications (GSM), a popular wireless telecommunication network standard. Such protocol integration is the result of EAP's extensibility, and thus significantly enhances the capability of serial authentication that enables greater security key strength through multiple authentication triplets in GSM.

3. *Remote Authentication Dial-In User Service* (*RADIUS*)

RADIUS[19] provides centralized access, authorization, and accounting (AAA). It defines an authentication server's function set and a protocol that facilitates message exchange between a network access server (NAS) and an authentication server. When used in an 802.11i WLAN setting, the RADIUS protocol is a request-response protocol that specifies the format and exchange procedure for authentication messages between an access point (AP) and an authentication server (AS). The message sent from AP to AS is called "RADIUS-Access-Request" which encapsulates the STA's EAP-Response/Identity message. Then, AS sends a "RADIUS-Access-Challenge" to AP which is further relayed to STA as an EAP-Request. After receiving the challenge, STA constructs a proper EAP-Response and AP delivers it to AS in RADIUS-Access-Request. Finally, based on STA's response, AS makes a decision about whether to grant or decline STA's request and replies with "RADIUS-Access-Accept" (success) or "RADIUS-Access-Reject" (fail).

While none of the IEEE 802.1X, EAP, and RADIUS was designed specifically for wireless mobile network systems, the integration of their functions fits right into the requirements of modern Wi-Fi network systems and applications. Therefore IEEE 802.11i and WPA2 adopt this security model. Since the publication of IEEE 802.11i, many new security features have been proposed and tested in WLAN-based wireless mesh networks. Those with significant security implications in wireless mesh networks are discussed in the following sections.

9.4.2 Authentication and Key Management

In wireless mesh networks, authentication is the process of establishing and confirming the identities of two or more participating entities which may include mesh stations, mesh access points, and mesh portals. Incorporated with the authentication, key management techniques are employed to create, exchange, and store cryptographic keys that may be used as initial identity credentials, intermediate secrets, and final authentication results.

1. *Simultaneous Authentication of Equals*

Adopted by the IEEE 802.11s standard, Simultaneous Authentication of Equals (SAE)[20] is a peer-to-peer authentication protocol for wireless mesh networks[2]. It employs password authenticated key exchange mechanism to provide resistance to passive attacks, active attacks, and dictionary attacks.

As a result, mutual authentication and a cryptographically strong shared secret key are established between the two peers.

SAE's approach is quite different from a traditional interconnected distributed system where communicating parties are classified as clients or servers depending on their roles. Message exchanges usually are initiated by the client (such as Internet browsing protocol HTTP). In this client-server paradigm, the associated security protocols (including authentication process) follow the client-server message exchange pattern and are not flexible with the role of protocol "initiator". SAE adopts the peer-to-peer model and does not differentiate between the roles of two entities. That is, either entity can start the process of security message exchange or both sides can start at the same time (the reason that it is called "simultaneous" authentication of "equals").

The effective operation of SAE needs a "finite cyclic group" which can be based on either "prime modulus groups" or "elliptic curve groups"[20]. The comparison of the essential features between these two approaches is shown in Table 9.2.

Table 9.2 Prime modulus groups vs. elliptic curve groups

	Prime modulus groups	Elliptic curve groups
Group based on	Exponentiation of integers modulo a prime	Elliptic curves over a finite field
Generate stronger shared key	Require larger prime	Require larger group
For a given key strength	Use a larger group size	Can use a smaller group size
Scalar operation (•)	Generator raised to a scalar power	A point on the curve multiplies by a scalar
Element operation (◇)	Two elements' modular multiplication	A point on the curve adds with another point on the curve
Inverse of a group element	Two elements are the inverse of each other if their product modulo the group prime is 1	Two points on the curve are the inverse of each other if their sum is the "point at infinity"

From the above comparison, it is clear that SAE can trade the execution speed for the key strength by adjusting the size of prime (for prime modulus groups) or the size of group (for elliptic curve groups) which directly controls the required computational resources for SAE.

The protocol exchange algorithms[20] using the above two groups are shown in Table 9.3.

The functions and notations in these two algorithms are as follows.

- A and B: participating entities.
- L: an ordering function determining the "greater" identity from input entities.
- SS: shared secret between A and B.
- H: a one-way "random oracle" function.
- |: a symbol for concatenation.
- KDF: a key derivation function that elongates the input string to the

specified length.
- PWE: password element.
- p: the prime of the curve (in elliptic curve groups) or the group prime (in prime modulus groups) with order r.
- len: the length of p.

Table 9.3 SAE algorithms: fixing password element

Prime Modulus Group	Elliptic Curve Group
if $L(A, B) = A$ then IDseq $= A\|B$	$i = 1$
else IDseq $= B\|A$	repeat
$n = H$ (IDseq | SS)	if $L(A, B) = A$ then IDseq $= A\|B$
$z =$ KDF (n, len) mod p	else IDseq $= B\|A$
PWE $= z^{((p-1)/r)}$ mod p	$n = H$ (IDseq | SS |i)
	$x =$ KDF (n, len) mod p
	solve for y with the curve equation and x
	if n is odd then $y = -y$
	PWE $= (x, y)$
	$i = i + 1$
	until PWE is on the curve

At the end of these algorithms, the password element is created. Then the two entities A and B will take individual actions as shown in Table 9.4.

Table 9.4 SAE algorithms: creating shared key

A's Actions	B's Actions
Pick random numbers rand_A and mask_A	Pick random numbers rand_B and mask_B
$\text{scal}_A = (\text{rand}_A + \text{mask}_A) \bmod r$	$\text{scal}_B = (\text{rand}_B + \text{mask}_B) \bmod r$
$\text{elem}_A = \text{inverse}(\text{mask}_A \cdot \text{PWE})$	$\text{elem}_B = \text{inverse}(\text{mask}_B \cdot \text{PWE})$
Send scal_A and elem_A to B	Send scal_B and elem_B to A
$K = \text{rand}_A \cdot (\text{scal}_B \cdot \text{PWE} \diamondsuit \text{elem}_B)$	$K = \text{rand}_B\ (\text{scal}_A \cdot \text{PWE} \diamondsuit \text{elem}_A)$
$\text{tok}_A = H(F(K)\|F\ (\text{elem}_A)\|\ \text{scal}_A\|F\ (\text{elem}_B)\|\ \text{scal}_B)$	$\text{tok}_B = H(F(K)\|F\ (\text{elem}_B)\|\ \text{scal}_B\|F\ (\text{elem}_A)\|\ \text{scal}_A)$
Send tok_A to B	Send tok_B to A
Verify tok_B (sent by B)	Verify tok_A (sent by A)
Shared key $= H(F(K)\|F\ (\text{elem}_A \diamondsuit \text{elem}_B)\|\ (\text{scal}_A + \text{scal}_B) \bmod r)$	Shared key $= H(F(K)\|F\ (\text{elem}_A \diamondsuit \text{elem}_B)\|\ (\text{scal}_A + \text{scal}_B) \bmod r)$

In Table 9.4, F is a bijective function with element-to-number mapping property, and *inverse* is the finite cyclic group inverse function. The elegance of SAE is demonstrated at the completion of both sides' actions when a strong shared secret is created from potentially weak user passwords.

2. *Efficient Key Establishment*

Many applications running on wireless mesh networks require real-time constraints to be met. For example, the popular Skype software is a voice over IP (VoIP) application that digitizes human natural analog voice signal, packetizes and delivers them to the other party. In such application, end-to-end delay is expected to be less than 50 $\sim$ 150 ms in order to have a smooth audio conversation (and/or video session). However, as indicated in a report[21], the full authentication process in an EAP-based 802.11X system takes about 1 000 ms which is far beyond most real-time applications' latency requirement. While this would not pose any problems for scenarios where re-authentications are not needed (i.e., authentication process is completed before the secure communication session starts), significant performance degradation may occur in a "multi-domain" wireless mesh network environment where mobility-based handoffs trigger essential re-authentications. Thus, a more efficient key establishment method is considered necessary in such conditions. In [22], the HMSF-AKES scheme is proposed to enable fast mutual authentication and pairwise key agreement between security entities in multi-domain wireless mesh networks.

HMSF-AKES is a hierarchical multivariable symmetric function (HMSF) based authenticated key establishment scheme (AKES). Its operation requires the completion of five steps.

(1) Individual Domain Function Generation. In each AAA server, a four variant two-level hierarchical domain function with a desired symmetric property is generated.

(2) Cooperative Federated Function Initialization and Distribution. Each AAA collects the generated domain functions from all participating AAAs, and uses them to compute the federated function.

(3) Individual Function Initialization and Distribution. Each AAA uses clients' registration IDs to evaluate the computed federated function.

(4) Authenticated Pairwise Master Key Generation. By exchanging the IDs of mesh security entities and their corresponding home domains, the pairwise master key can be computed.

(5) Pairwise Session Key Generation. Based on the obtained pairwise master key, more pairwise session keys can be derived for future communication protection.

3. *Channel Probing for Shared Key Generation*

Conventional Diffie-Hellman key exchange has been proven to be secure and effective and broadly adopted in symmetric key cryptography systems to establish a shared secret key between two security entities. The new channel probing technique[23] is proposed based on the assumption that in the future the realization of quantum computing can break Diffie-Hellman protocol in reasonable amount of time. The channel probing itself is, however, immune from those attacks based on immense computation power (which is the key feature of quantum computing).

Channel probing is a process to gather parameter information from the wireless channel between the two communicating parties. The most widely used channel parameter for this purpose is the received signal strength (RSS). It is assumed that the target wireless fading channel exhibits reciprocal and location-specific properties so that the two parties can collect highly correlated channel information and generate identical shared secret keys. How secure is this new type of system? It is shown that as long as the eavesdroppers are located from the rightful key owners for more than a half of the radio wavelength (i.e., $\lambda/2$), they would not be able to create the same keys because the channel information would be significantly different at a short distance away.

9.4.3 Path Selection and Routing

Wireless mesh networks and mobile ad hoc networks have many similarities from the perspective of routing (or path selection). Both types of network employ multi-hop routing strategy to deliver packets from one node to another (or to others in the multicast or broadcast cases). Therefore, it is not surprising that many MANET routing security mechanisms can be applied to WMNs. Most secure routing protocols deal with external threats and attacks. That is, adversary nodes are not assumed to be able to gain full control of legitimate nodes (those that have been authenticated). To defend against such external threats and attacks, there are three basic categories of approach for secure routing. They are based on asymmetric cryptography, symmetric cryptography, and the hybrid of the two[9]. Among many options, the most relevant secure routing protocols are as follows.

- *Authenticated routing for Ad Hoc networks (ARAN) protocol*[24]. ARAN protocol utilizes the digital certificates in asymmetric cryptography systems. In ARAN, routing messages are cryptographically protected by digitally signing (with private-key) the attached public-key certificate. Since the certificate itself is signed by a trusted certificate authority (CA), its integrity is assumed. Based on this, the receiver node can verify the legitimacy of the received routing messages and thus thwart routing information fabrication.
- *Secure Efficient Ad Hoc Distance (SEAD) vector routing protocol*[25]. SEAD protocol is a secure routing protocol based on the design of the Destination Sequenced Distance Vector (DSDV) routing protocol. The main concept in SEAD is to employ the hash chain technique to protect important routing information (such as the sequence number and hop count). Hash chain are effective against adversaries because of their one-way property that makes the derivation from output back to input basically impossible (computationally too expensive). In comparison, computing hash result is less time and resource consuming than the asymmetric cryptography approach. However, SEAD requires synchronized clocks.
- *Secure Ad Hoc on demand distance vector (SAODV) protocol*[26]. SAODV protocol is a hybrid approach to take advantage of the positives from both asymmetric and symmetric cryptography systems. Since only some of the fields (i.e., mutable fields) in routing messages could change in the routing process, SAODV use the economical and light-weight hash chain to protect mutable fields.

9.5 WMN Security Standards and Products

The promising wireless mesh network and security standard, relevant commercial products, and an important project are discussed in this section.

1. *WMN Standard*

So far the most important standardization process for Wi-Fi-based wireless mesh network is the IEEE 802.11s. However, at the time of this writing, it is still in the draft development phase (i.e., not a standard yet). The latest status can be found at the task group's website at http://www.ieee802.org/11/Reports/tgs_update.htm.

From the security viewpoint, the Authentication of Equals (SAE) is most likely to be included in the finalized specification. Besides, other security mechanisms defined in IEEE 802.11i are ready to be deployed for many security services that are necessary for the robust and secure operation of WMNs.

2. *WMN Products*

There are many companies that have developed wireless mesh network related products and solutions. Due to proprietary information, a full scale product performance comparison is difficult to conduct. In terms of WMN security characteristics, SANS Institute compares the following four commercial products: Tropos 5120, Cisco AP1500, Motorola HotZone Duo, and Proxim 4000M[27]. The results show that all four of them have IEEE 802.11i/WPA2 client access and multiple VLAN/SSID security policies. While the support for device authentication is provided, it is achieved differently. Cisco AP1500 and Motorola HotZone Duo use X.509v3, Tropos 5120 is WPA-PSK, and Proxim 4000M uses a simple shared key. All four products provide inter-mesh AP payload encryption through 128-bit AES, and secure management through HTTPS and SNMPv3. The most significant difference is that only Tropos 5120 is capable of mesh protocol integrity protection, while the other three are not.

3. *OPEN80211s*

Open80211s is a project to closely monitor the standardization progress of IEEE 802.11s and implement its functions faithfully in the open source Linux operating system[2]. The way it is integrated into the Linux kernel is demonstrated in Fig. 9.4.

As cited on its website homepage[28], "open80211s is a consortium of companies who are sponsoring (and collaborating in) the creation of an open-source implementation of the emerging IEEE 802.11s wireless mesh standard. The resulting software will run on Linux on commodity PC hardware." With this vision, the consortium has set its ambitious goal: Based on the IEEE 802.11s draft/standard, open80211s aims to provide the first open source implementation that can be used, understood, and contributed by anyone who is interested. Ultimately, it hopes to develop a large wireless mesh network

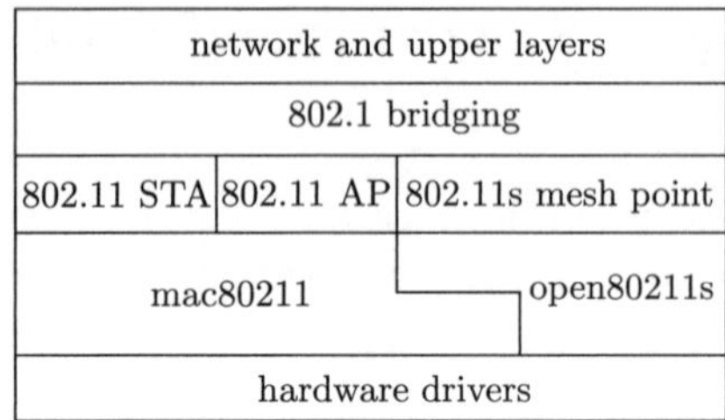

Fig. 9.4 Open80211s in Linux.

that connects all Linux wireless computers and devices around the world.

While there are many designs and proposals for wireless mesh network security features, the main security module that has been implemented in open80211s is the Simultaneous Authentication of Equals (SAE), which is the reason that SAE was introduced in detail in this chapter.

9.6 Conclusion

With its attractive economic and flexible factors, wireless mesh network technology is positioned to take over a large chunk of the telecommunication and data communication market. With such power comes great responsibility in securing all data generated and passed through the WMNs. This chapter first provided a short introduction on the WMN and its distinct characteristics in comparison with cellular networks, Internet, and mobile ad hoc networks. The WMN security challenges, potential threats, and attacks were discussed. With WMN vulnerabilities in mind, most relevant security mechanisms that can be utilized to deter threats and attacks were analyzed. With the most promising one likely to be Simultaneous Authentication of Equals (SAE) which has already been included in the IEEE 802.11s standardization process. Finally, some commercial WMN products were compared in terms of their security functions. An open source project named "open80211s" is aiming to connect all Linux-based wireless computers and devices. It is worth noting that there is no single silver bullet which can solve all security issues in WMNs. It will take the right combination of policies, mechanisms, services, and executions to furnish robust and secure wireless mesh networks.

References

[1] Faccin SM, Wijting C, Kenckt J, Damle A (2006) Mesh WLAN networks: concept and system design. IEEE Wireless Communications, 13(2): 10–17.

[2] Hiertz G R, Denteneer D, Max S, Taori R, Cardona J, Berlemann L, Walke

B (2010) IEEE 802.11s: The WLAN Mesh Standard. IEEE Wireless Communications, 17(1): 104 – 111.

[3] Kraemer K L, Dedrick J, Sharma P (2011) One Laptop Per Child (OLPC): A Novel Computerization Movement. Proceedings of the 44th Hawaii International Conference on System Sciences, pp. 1 – 10.

[4] Yarali A, Ahsant B, Rahman S (2009) Wireless Mesh Networking: A Key Solution for Emergency & Rural Applications. The Second International Conference on Advances in Mesh Networks (MESH 2009), pp. 143 – 149.

[5] IEEE P802.11 — Task Group S (2011) Status of Project IEEE 802.11s Mesh Networking. http://www.ieee802.org/11/Reports/tgs_update.htm. Accessed 30 June, 2011.

[6] Carrano R C, Magalhães LCS, Saade DCM, Albuquerque CVN (2011) IEEE 802.11s Multihop MAC: A Tutorial. IEEE Communications Surveys & Tutorials, 13(1).

[7] Salem N B, Hubaux J P (2006) Securing wireless mesh networks. IEEE Wireless Communications, 13(2): 50 – 55.

[8] Glass S, Portmann M, Muthukkumarasamy V (2008) Securing Wireless Mesh Networks. IEEE Internet Computing, 12(4): 30 – 36.

[9] Zhang W, Wang Z, Das S K, Hassan M (2007) Security Issues in Wireless Mesh Networks. Wireless Mesh Networks, pp. 309 – 330.

[10] IEEE 802.11 Working Group (2007) IEEE Standard for Information Technology — Telecommunications and Information Exchange between Systems — Local and Metropolitan Area Networks — Specific Requirements — Part 11: Wireless Medium Access Control (MAC) and Physical Layer (PHY) Specifications.

[11] Wi-Fi Alliance (2011) http://www.wi-fi.org/. Accessed 30 June, 2011.

[12] Edney J, Arbaugh W A (2004) Real 802.11 Security: Wi-Fi Protected Access and 802.11i. Addison-Wesley Professional, Boston.

[13] Frankel S, Eydt B, Owens L, Scarfone K (2007) Establishing Wireless Robust Security Networks: A Guide to IEEE 802.11i. Recommendations of the National Institute of Standards and Technology, NIST Special Publication 800-97.

[14] Kurose J F, Ross K W (2010) Computer Networking: A Top-Down Approach (5th edition). Wesley, New York.

[15] IEEE Standard for Local and metropolitan area networks (2004) 802.1X: Port - Based Network Access Control.

[16] Aboba B, Blunk L, Vollbrecht J, Carlson J, Levkowetz H (2004) IETF RFC 3748: Extensible Authentication Protocol (EAP).

[17] Simon D, Aboba B, Hurst R (2008) IETF RFC 5216: The EAP-TLS Authentication Protocol.

[18] Haverinen H, Salowey J (2006) IETF RFC 4186: Extensible Authentication Protocol Method for Global System for Mobile Communications (GSM) Subscriber Identity Modules (EAP-SIM).

[19] Rigney C, Willens S, Rubens A, Simpson W (2000) IETF RFC2865: Remote Authentication Dial-In User Service (RADIUS).

[20] Harkins D (2008) Simultaneous Authentication of Equals: A Secure, Password-Based Key Exchange for Mesh Networks. In Proceedings of the Second International Conference on Sensor Technologies and Applications (SENSORCOMM '08), pp. 839 – 844.

[21] Aboba B (2003) Fast Handoff Issues. IEEE 802.11 Working Group, IEEE-03-155r0-I.

[22] He B, Joshi S, Agrawal D, Sun D (2010) An efficient authenticated key establishment scheme for wireless mesh networks. IEEE Globecom Ad-hoc and Sensor Networking Symposium (GC10 - AHSN).

[23] Wei Y, Zeng K, Mohapatra P (2011) Adaptive Wireless Channel Probing for Shared Key Generation. IEEE Infocom 2011.

[24] Sanzgiri K, Dahill B, Levine B, Shields C, Belding Royer E M (2002) A secure routing protocol for ad hoc networks. In Proceedings of 2002 IEEE International Conference on Network Protocols (ICNP), pp. 78 – 87.

[25] Hu Y C, Johnson D B, Perrig A (2003) SEAD: Secure efficient distance vector routing for mobile wireless ad hoc networks. Ad Hoc Networks, pp. 175 – 192.

[26] Zapata M G, Asokan N (2002) Securing ad hoc routing protocols. In Proceedings of the 2002 ACM Workshop on Wireless Security (WiSe 2002), pp. 1 – 10.

[27] Gerkis A (2006) A Survey of Wireless Mesh Networking Security Technology and Threats. http://www.sans.org/reading_room/whitepapers/honors/survey-wireless-mesh-networking-security-technology-threats_1657. Accessed 9 December, 2010.

[28] open80211s (2011) http://open80211s.org/. Accessed 30 June, 2011.

Chapter 10
Security in RFID Networks and Communications

Chiu C. Tan[1] and Jie Wu[2]

Abstract

Radio frequency identification (RFID) networks are an emerging type of network that is posed to play an important role in the Internet-of-Things (IoT). One of the most critical issues facing RFID networks is that of security. Unlike conventional networks, RFID networks are characterized by the use of computationally weak RFID tags. These tags come with even more stringent resource constraints than the sensors used in sensor networks. In this chapter, we study the security aspects of RFID networks and communications. We begin by introducing the main security threats, followed by a discussion of various security mechanisms used to protect RFID networks. We conclude by studying the security mechanism of an actual large scale RFID deployment.

10.1 Introduction

Radio frequency identification (RFID) technology consists of small inexpensive computational devices with wireless communication capabilities. Currently, the main application of RFID technology is in inventory control and supply chain management fields. In these areas, RFID tags are used to tag and track physical goods. Within this context, RFID can be considered a replacement for barcodes.

RFID technology is superior to barcodes in two aspects. Firstly, RFID tags can store more information than barcodes. Unlike a barcode, the RFID tag, being a computational device, can be designed to process rather than

1 Temple University, Philadelphia, PA, USA. E-mail: cctan@temple.edu.
2 Temple University, Philadelphia, PA, USA. E-mail: jjewu@temple.edu.

just store data. Secondly, barcodes communicate using an optical channel, which require the careful positioning of the reading device with no obstacles in-between. RFID uses a wireless channel for communication, and can be read without line-of-sight, increasing the read efficiency.

The pervasiveness of RFID technology in our everyday lives has led to concerns over whether these RFID tags pose any security risk. For example, consider an RFID tag affixed to clothing, this type of tag contains information such as the brand and model of the clothing. This type of information is used for inventory purposes. A thief armed with an RFID reader can, however, use the same information to select wealthy targets, which are more likely to wear more expensive clothes, to pickpocket.

The future applications of RFID make the security of RFID networks and communications even more important than before. The ubiquity of RFID technology has made it an important component in the Internet-of-Things (IoT), a future generation Internet that seeks to mesh the physical world together with the cyber world[1]. RFID is used within the IoT as a means of identifying physical objects. For example, by attaching an RFID tag to medication bottles, we can design an RFID network to monitor whether patients have taken their medications. RFID readers can be used to determine when medication bottles have been removed from the medicine cabinet, this information can be combined with additional information, such as weight sensors that record the weight of medicine bottle, to infer whether a patient has taken his medication. Such applications, while undoubtedly useful, opens the door to allow malicious entities to launch attacks like determining what types of medication a person is taking.

Given the stakes, it is unsurprising that RFID security has attracted the attention of researchers. In recent years, there have been numerous RFID security protocols proposed, and new RFID vulnerabilities discovered. The difficulty in securing RFID lies in the resource constraints of the RFID tags, which makes it impossible to adopt existing security solutions from other fields such as mobile computing or wireless networking, onto RFID networks.

This chapter studies the security of RFID networks. Firstly, we discuss some background on RFID networks, followed by an introduction to main RFID threats. We then review and analyze some basic RFID security protocols, followed by a discussion on more advance attacks and defense. Finally, we discuss the security of industry standard RFID protocols.

10.2 RFID Network Primer

An RFID network consists of three basic components: RFID tags, RFID readers, and backend servers. In an RFID network, each RFID tag contains small amounts of information which are affixed to physical objects. RFID readers read the information from these tags as the physical object moves around a given area. The information is then transmitted from the readers to backend servers for processing to service higher level applications. Fig. 10.1 shows the interactions between the three components.

Fig. 10.1 shows that all interactions are reader driven. The RFID tag never initiates any communications. The RFID reader can be configured like a Wi-Fi access point (AP) beaconing to periodically broadcast a query to read tags in the vicinity, or the query can be manually triggered. The communication channel between the RFID reader and the backend server can be either wired or wireless, and is assumed to be secure. We also assume that some access control policy is in place to regulate reader access to the backend server. The channel between the reader and the tag is assumed to be insecure. The majority of RFID security research is focused on securing this wireless channel.

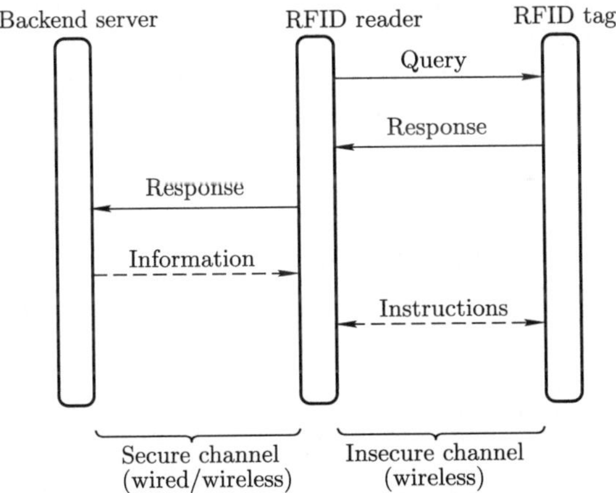

Fig. 10.1 Basic interaction between the components. Dashed line indicates optional operations. There can be multiple interactions between the reader and tag in the "Instructions" command, as denoted using double headed arrows.

The two optional operations, shown the Fig. 10.1, are generally used when the reader needs to write any information onto the RFID tag. To protect the integrity of the RFID tag data, writing to the tag's memory typically requires some sort of password which is stored in the backend servers.

An example of an RFID network is an RFID-enabled hospital. Patients are given a unique RFID tag to wear. The tag contains the patient's unique ID. RFID readers installed throughout the hospital can track the movement of patients through the reading of the tag IDs. In addition, medical treatments (e.g., blood bags, pills, etc.) are also embedded in RFID tags. The backend servers will associate a patient's RFID tag ID with the appropriate treatments, and a nurse will scan all the tags before administrating treatments.

10.2.1 RFID reader characteristics

The RFID network may consist of both mobile and static RFID readers. The mobile reader combines a processing unit and antenna together, and resembles a smartphone type device. The processing unit is used to communicate with the backend server and issue commands to the antenna. The antenna is used to broadcast and receive messages. A user will aim the reader at a set of RFID tags to query them. Information from the mobile reader can be transmitted to the backend servers wirelessly. A static RFID reader has the antenna permanently positioned at a specific location (i.e. entrance to a specific hallway). Multiple antennas may share a single processing unit, and these antennas are connected to the processing unit via wired channel.

We usually do not make a distinction between the antenna and the processing unit in RFID security literature. They are both simply referred as "RFID reader". However, this distinction can be important for performance reasons where angles of the antennas matter.

The purpose of the RFID reader is to communicate with RFID tags, and send the information back to the backend servers. Besides, the reader is responsible for regulating tag responses. One of the limitations of the RFID tag is that the tag cannot perform carrier sensing. Instead, the RFID reader acts as a coordinator to regulate tag communications. Most RFID security protocols however ignore this function and simply assume that there is only a single reader querying a single tag.

10.2.2 RFID tag characteristics

Each RFID tag contains a unique identifier (id). Once a tag is affixed to a physical object, the id becomes a representation of that object.

1. *Types of RFID tags*

There are three general types of RFID tags, active, semi-active, and passive RFID tags.

- *Active RFID tags.* This type of RFID tag contains an internal battery which is used to let the tag perform more complex operations, such as monitor temperature, as well as boost the communication with an RFID reader. The communication range of an active tag can be over 100 meters. An active tag is the most powerful type of RFID tag, and is also the most expensive.
- *Semi-active RFID tags.* This type of tag also contains an internal battery, but unlike an active tag, the battery is only used for the tag's internal operations, and not for communication. A semi-active RFID tag relies on RFID reader to supply the necessary power for communication. Note that semi-active tags are sometimes known as semi-passive tags.

- *Passive RFID tags.* This type of RFID tag have the lowest cost (pennies per tag), and unsurprisingly, are the most prevalent type of RFID tags. A passive tag has no internal batteries, and relies on the RFID reader to supply the power needed to perform all tag operations and communication. In the rest of this chapter, our focus is on this type of tags.

2. *Communication range*

The conventional range of the tag can range from several centimeters, for RFID tags operating in the 13.56 Hz, to over a dozen meters for RFID tags operating in the 902-928 MHz. Due to the physical characteristics of the reader and tag, the signal being passed from the reader to the tag is stronger than that from the tag to the reader. This means that for certain operations like eavesdropping, it will be easier to hear the RFID reader's commands than it is for the tag's response.

In terms of security, however, we cannot rely on the conventional communication range. Determining the RFID tag communication range for security purposes is difficult for two reasons. Firstly, RFID tag responses are sensitive to environmental conditions. Reading an RFID tag on credit card in a purse placed in a handbag is very different from reading a tag placed on a store shelf. Secondly, when launching an attack, the adversary can use non-standard equipment that is more powerful than regulation equipment. There have been experiments on querying RFID tags in "realistic" environments (tag placed in a person's wallet), but such experiments are limited by the use of conventional equipment[2].

3. *Computational ability*

Despite having no battery power, passive RFID tags do exhibit a wide range of capabilities. RFID tags contain limited amounts of persistent storage capacity, and the storage on a tag can be read-only, write-once, or multiple writes. The difference between a read-only and a write-once tag is that for a read-only tag, the initial information is usually stored when the tag is manufactured and not transmitted by a RFID reader. The distinction between tags that support multiple writes and those that do not is important for RFID security, since some protocols require authorized readers to change the stored data after every successful read. Current commercial RFID tags can perform functions such as matching bit strings, exclusive-ORs, generate random numbers, cryptographic hash functions, and symmetric key operations. Within RFID security research, there is work on designing security solutions that do not use these functions. For instance, there are protocols that use only hash functions, or do not use random number generators. The reason is that engineering more functions will increase the cost of the tag, and using weaker tags are cheaper. Table 10.1 lists the capabilities of a sample RFID tag.

Table 10.1 Sample RFID tag security capability

Type of tag	Security capabilities
Low end	32-bit access and kill password. The kill password is used to render the RFID tag non-responsive to further RFID reader queries. 64-bit fixed ID value. ID assigned at time of manufacture, and cannot be changed. Used for certain counterfeit tracking operations.
High end	64-bit mutual authentication protocol (proprietary). Stream encryption capabilities (proprietary). Support multiple passwords for fine grain access control. Different memory locations can require different passwords to access.

1 Based on EPC Class 1-Gen 2 standards and Alien Technology ALN-9640 tags.
2 Based on Atmel ATA6286 Crypto RF tag.

10.3 Security Requirements

The challenges in security RFID networks lie in securing the operations involving RFID tags. This is because the severe resource limitations of tags make it difficult to implement conventional security mechanisms. RFID readers and backend servers on the other hand, can be secured using existing security techniques. In this section, we begin by examining the key RFID security requirements, followed by more specific requirements for certain RFID applications.

There are three key RFID security requirements: prevent unauthorized access, prevent illicit tracking, and prevent or detect skimming. These form the basic requirements for most RFID applications.

- *Prevent unauthorized access.* There are two ways which unauthorized access can occur. The first is when an unauthorized RFID reader queries and obtains usable information such as the tag ID from the RFID tag. RFID tag design requires the tag to respond to any query. Any reader can query the tag and get a response. Preventing unauthorized access refers to allowing only authorized readers to obtain usable information. The second way which unauthorized access can occur is via eavesdropping. An adversary obtains usable information by observing the over-the-air communications between a legitimate reader and a tag.
- *Prevent illicit tracking.* This requirement addresses one of the main privacy concerns over the use of RFID technology. Illicit tracking exploits the fact that RFID tags always respond to reader's query. An adversary that queries and obtains the same tag response at multiple locations can infer that the same tag has visited those locations. Since RFID tags are affixed to physical objects, for instance clothing, this implies that the same person has visited those locations. Note that satisfying the first requirement does not automatically satisfy this requirement. A tag that returns a constant, encrypted response will prevent unauthorized access, since the adversary cannot determine the tag contents. However, the constant ciphertext can be used to perform illicit tracking.
- *Prevent or detect skimming.* Skimming is an attack whereby the adversary

observes the interactions between a legitimate RFID reader and a tag, and tries to create a fake RFID tag that mimics a real one. The adversary succeeds when his fake tag can pass off as a real tag. Skimming is a concern when RFID is used to authenticate documents such as driver licenses or passports. For instance, an adversary that tries to create a fake drivers license may attempt to observe the interactions of an RFID tag embedded in a legitimate drivers license to create his fake RFID tag. Generally, the adversary performing a skimming attack does not have physical access to the RFID tag.

In addition to the key requirements listed above, there are more specific security requirements that are important for certain applications. Applications that transfer ownership of the tag, either temporarily or permanently will require that the previous owner of the RFID tag can no longer access the data stored in the tag. This requirement is known as secure ownership transfer. A related requirement is forward-security. This requirement means that an adversary that learns of an RFID tag secret, for instance, by physically compromising the tag, cannot determine previously encrypted information from that tag. Secure RFID search is used when a user wishes to locate a particular tag from a large collection of RFID tags. The requirement of a privacy-preserving RFID search is to ensure that searching does not leak information about the RFID tag.

There are two advance requirements that cannot be easily solved using the solutions address on the basic requirements.

- *Defend against Mafia fraud.* This is a relay-type attack where the adversary deploys a fraudulent reader and tag. The fraudulent reader will query the real RFID tag, and then relay the information to the fraudulent tag to replay to a legitimate reader. Defense against this type of attack is needed for applications that use RFID tags for access control purposes (e.g. opening a car door), or for payment applications like credit cards.
- *Grouping proofs.* A grouping proof requires the RFID reader to prove that a set of RFID tags were read at the same time and location. For instance, a patient may be required to take three types of medications at the same time. The nurse with a mobile RFID reader can generate a grouping proof that captures all three RFID tags (affixed to the medication containers) were present at the same time to prove that the patient was correctly medicated.

10.4 Hardware Based Solutions

A straightforward solution to provide security is to physically disable the RFID tags. The idea of a "clipped tag" was proposed where the RFID tag was designed to allow the user to separate the RFID chip (contain the tag data) from the tag antenna (used to power the chip)[3]. This way, no RFID readers can query the tag, and thus making it secure. Later work improved upon this idea by allow the clipped tag to continue to be read by an RFID reader, but at a much shorter distance[4]. This approach resolves the key RFID security requirements by forcing adversary to be physically very close to the RFID tag to read any data, which makes such attacks easily detectable.

An important argument against disabling the RFID tag is that the process is irreversible. Instead, an alternative is to design a special device to disrupt the RFID operations, which a user can carry with them. This idea was first proposed by Juels et al. in the form of a blocker tag, a special RFID tag which can be programmed to block certain tag IDs that the user considers sensitive[5]. The blocker tag is also a passive RFID tag. Feldhofer et al. proposed a watchdog type device to alert users when an RFID reader is querying their tags[6]. Later work by reference [7] developed a more powerful battery operated device, the RFID Guardian, that intercepts the RFID reader's signal and only allow signals from authorized readers to reach the tag. Since the adversary never gets any response from the RFID tag, the guardian provides the needed security requirements.

Hardware based solutions, while being an important component in RFID security research, are less popular than protocol based solutions. There are several possible reasons for this. Firstly, hardware type solutions tend to be more expensive due to the use of external devices. Secondly, such solutions can potentially disrupt operations of other RFID tags belonging to other users, which make it more difficult to gain acceptance. Finally, when RFID tags were initially deployed, there were concerns that tag manufacturers may be unwilling to engineer security protections into the tag since this will increase their manufacturing cost. Hardware based solutions are practical in that context since they do not rely on the tag manufacturers. In recent years however, public awareness over security RFID appear to have led to the deployment of more secure RFID tags, making hardware based solutions less attractive.

10.5 Basic Protocol Based Solutions

Protocol based RFID solutions rely on the RFID tags performing certain operations to provide the key security requirements to prevent unauthorized access, tracking, and skimming.

10.5.1 Different RFID Protocols

There are too many RFID protocols in the literature to be included in this chapter. Instead, we attempt to categorize them based on the focuses of these protocols, and highlight just a few works in each category. We have elected to avoid discussing RFID protocols designed from specialized applications such as banknotes[8] or supply chains[9]. A good resource for the latest updates can be found in reference [10].

1. *Improving backend performance*

One approach lies in improving the performance of the backend server. From Fig. 10.2, we see that the backend server needs to try all $(s : id)$ pairs to determine the correct secret s to use in order to obtain the tag *id*. The reason that the RFID tag does not inform the backend server which secret s to use is to defend against illicit tracking. As a result, the RFID tag has to output a different random value each time it is queried. A more detailed analysis of protocols designed to alleviate the bottleneck can be found in reference [11].

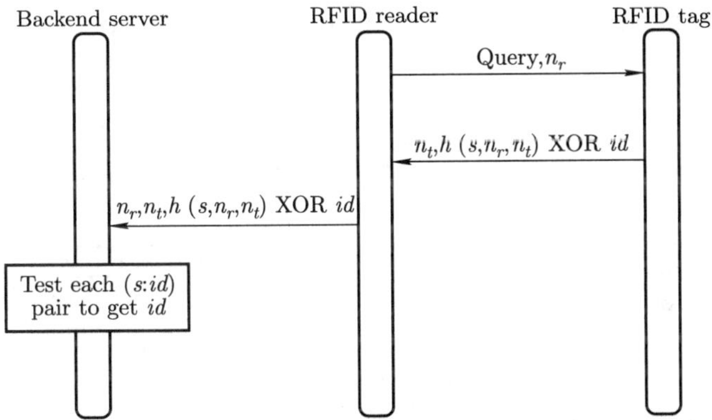

Fig. 10.2 Basic protocol that defends against key RFID security requirements. Modified from Tan et al.[22]. Random numbers from the reader and the tag are denoted as n_r and n_t respectively. The variables s and *id* denote the RFID tag's secret and *id*. Each tag has a unique secret and *id* that is assigned by the backend server. A conventional hash function is denoted as $h()$.

One example of such protocols is a time-based solution proposed by reference [12]. The intuition is to let the backend server maintain a lookup table associated with the tag secret that is hashed with a timestamp, $(h(s, t) : id)$. The backend server can pre-compute this table each time t. Each time the reader queries the tag, the reader will send the timestamp t, and the tag will respond with $h(s, t)$. This way, the backend server can obtain the corresponding *id* immediately using the lookup table. Later work by reference [13] and reference [14] improves on this approach.

2. *Using lightweight primitives*

In Fig. 10.2, the RFID tag uses a hash function $h()$ to protect its response. Given the hardware limitations of the RFID tag, an area of RFID research attempts to design solutions that do not use hash functions or symmetric keys to provide security. One popular approach is generally known as HB family of protocols which is after the authors[15]. The HB family of protocols uses scalar products and exclusive-ORs to design their protocols. Work by reference [16] first proposed HB protocols that defend against different RFID attacks. A general survey of the HB family can be found in reference [17].

3. *Generating random numbers*

RFID protocols make extensive use of random numbers. A weak source of random numbers will allow the adversary to launch tracking. The use of random numbers to defend against tracking depends on the quality of the random number generated by the weak RFID tag. Work by Holcom et al.[18], J. Melia et al.[19], and Peris et al.[20] explores this problem in further detail.

10.5.2 A Detailed Look at a Simple RFID Protocol

Here we introduce a protocol modified from Tan et al.[21] to illustrate how a protocol based solution provides key RFID security requirements. Table 10.2 lists the notation used in this chapter. Fig. 10.2 illustrates the protocol.

Table 10.2 Notations used

n_t	Random number generated by RFID tag
n_r	Random number generated by RFID reader
s	RFID tag secret
id	RFID tag id
$h()$	Cryptographic hash function
t	Timestamp

From the protocol shown in Fig. 10.2, we see that when the reader queries the tag, the reader will first transmit a random number, n_r, to the tag. The RFID tag will respond by first generating its own random number, n_t, and then compute a response to protect its tag id using $h(s, n_r, n_t)$ XOR id. The reader then re-directs the tag's response, together with the random number it chose, to the backend server.

The role of the backend server is to determine the id of the RFID tag. Since the backend server is responsible for all the RFID tags, it maintains a list of all tag secret to tag id pairs $(s : id)$. Upon receive the message for the RFID reader, the backend server will know the two random numbers n_t and n_r chosen by the tag and reader respectively. From the list of $(s : id)$ pairs, the backend server will hash the secret s to generate $h(s, n_r, n_t)$ and XOR that against the response by the RFID reader, i.e. $h(s, n_r, n_t)$ XOR $h(s, n_r, n_t)$ XOR id. If the result matches the id in the list, the backend server will have determined the tag id. Otherwise, the backend server will continue to the next pair.

10.5.3 Security analysis

Here we will analyze how the basic protocol in Fig. 10.2 meets the key RFID security requirements.

The first requirement is to prevent unauthorized access to the RFID tag information. We first consider an unauthorized reader querying the tag. The adversary will issue its own random number n_r, and receive n_t, $h(s, n_r, n_t)$ XOR *id* from the tag. Since the backend server will not respond to the adversary, the adversary now has to determine *id* without any help from the backend server. The adversary succeeds if he is able to determine *id* from n_t, $h(s, n_r, n_t)$ XOR *id*. In order to get back *id*, we need to XOR with $h(s, n_r, n_t)$ using the correct s value, but the adversary only knows n_r and n_t, and not s. Thus, the adversary is unable to obtain *id*. Since the protocol uses a conventional hash function such as SHA, the adversary cannot obtain s from $h(s, n_r, n_t)$. The adversary can attempt to guess the value of s, but this can be defended against by using large enough values of s.

Another method of unauthorized access is for the adversary to be eavesdropping when a legitimate reader is querying a tag. Since the wireless channel between the reader and tag is assumed to be insecure, the adversary is able to learn n_r, n_t, and $h(s, n_r, n_t)$ XOR *id*. These pieces of information are similar to that obtained when the adversary queries the tag directly, which yields no useful information to the adversary.

The second requirement is to prevent illicit tracking. In this attack, the adversary needs to determine whether two tag responses belong to the same RFID tag. From Fig. 10.2, we see that the RFID tag has two pieces of information that remains constant, the tag *id* and tag secret s. However, each time the tag replies to a query, the tag will select a different random number, n_t, and thus, the resulting $h(s, n_r, n_t)$ XOR *id* will always be different for every response. This prevents any illicit tracking, since the adversary is unable to determine whether two responses are from the same RFID tag or not. This defense remains valid even if the adversary can select its own n_r value.

The third key requirement is to prevent or detect skimming. The adversary launching a skimming attack will observing the responses of a real RFID tag in attempt to create a fake tag that can pass off as a real tag. In the basic protocol, the adversary is able to observe the return value of n_t, $h(s, n_r, n_t)$ XOR *id*, However, it is unable to learn s or *id* based on the response. The adversary thus can only store $h(s, n_r, n_t)$ XOR *id* directly into a fake RFID tag. This skimming attack will be detected when a legitimate reader queries the RFID tag. The legitimate reader will issue its own random number, which we denote as n'_r to distinguish from the earlier n_r observed by the adversary. Since the fake tag does not know s or *id*, the fake tag can only return $h(s, n_r, n_t)$ XOR *id*, and not the correct $h(s, n'_r, n_t)$ XOR *id*. Since the backend server will attempt to test using n'_r and not n_r, this leads the backend server unable to find a correct (s, id) pair. Thus the skimming attack is detected.

10.6 Advance Protocol Based Solutions

Beyond the key RFID security requirements, there are some other RFID security requirements. This section discusses some protocols that address these requirements. Note that the protocols presented here may not necessary meet all the key security requirements because these advance protocols are generally designed to address specific issues or applications.

10.6.1 Defending against Mafia fraud

The mafia fraud has emerged as a challenging problem for RFID applications. This type of attack cannot be defended by the protocols mentioned earlier because a legitimate RFID reader is accessing data from a legitimate RFID tag. In other words, this type of attack can still work even if both the reader and the tag authenticate each other. This is illustrated using the basic protocol shown in Fig. 10.2.

Consider an application which uses RFID tag to open a door. The RFID reader will first read the tag and then transmit the information to the backend server. Once the backend server verifies the tag is legitimate, the door will open. To launch a mafia fraud attack, the adversary will first be in close proximity with a person holding a legitimate RFID tag. We refer to this person as the target. The adversary's accomplice will be standing near to the door. When the legitimate RFID reader issues a query, the adversary's accomplice will relay this message to the adversary, who will in turn issue it to the target's RFID tag. The target's RFID tag will respond to the adversary, who will then relay this back to his accomplice to transmit to the RFID reader. Since the RFID reader obtains the response from a legitimate RFID tag, the door will open and the adversary can gain access. Therefore the choice of protocol does not defend against this type of attack.

The intuition behind the defending against a mafia fraud is to accept an RFID tag's response if it is both valid and timely. Since the wireless transmission speed, the RFID tag computational time, and distance between the reader and tag are known, the RFID reader can estimate the amount of time needed to receive a response. If the arrival of the RFID tag response is late, the reader can deduce the distance travelled is longer than what is allowed, and thus reject the tag answer.

One of the main solutions against the mafia fraud is from Hancke et al.[23] and is shown in Fig. 10.3. We assume the system will define a maximum distance d, over which the reader is not suppose to authenticate a tag. In the protocol, we see that both reader and tag exchange random numbers. Assuming that the reader knows the tag secret s, both entities can compute $h(s, n_r, n_t)$. The tag will split this result into two queues, X and Y. At the same time, reader than generates a k bit challenge, $C_1, \cdots, C_k$, where k is a system defined parameter. The idea is that the reader will challenge the tag by sending over a bit C_i. If the C_i is 0, the tag will set A_i to the bit from queue X, and vice versa. The reader will keep the time it takes from sending C_i and receiving the A_i. A legitimate RFID tag that is within the approved distance will respond within the allocated time limit. A legitimate RFID tag that is further away will take a longer time to respond, due to the longer distance travelled, and thus is detected. More recent work on this topic can be found in reference [24] and reference [25]. An interesting idea of doing distance bounding for a group of RFID tags instead of just two tags has been proposed by Capkun et al.[26].

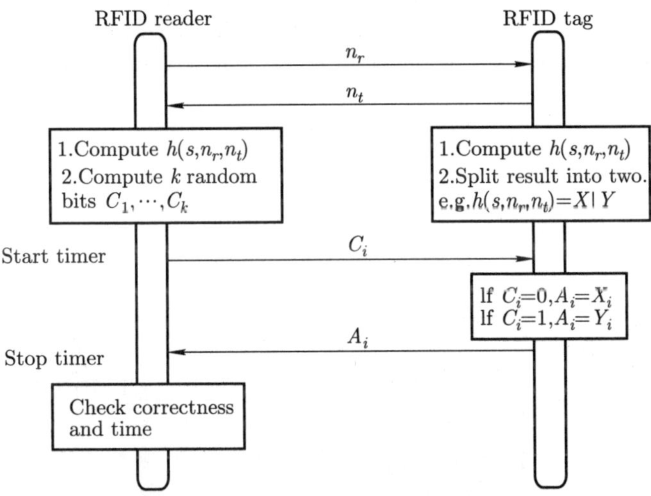

Fig. 10.3 Distance bounding protocol from Hancke et al. We assume that the reader already knows the tag secret s. We retain the notation from Table 10.2. X and Y are the bitstring resulting from dividing $h(s, n_r, n_t)$ into two. The protocol will repeat itself until C_k is transmitted from the reader to the tag. The variable k is a system defined parameter.

10.6.2 Grouping proofs

Grouping proofs are required for a reader to prove to the backend server that a set of RFID tags are physically close to each other. This type of proof typically requires a more advanced RFID tag that is able to maintain an atomic counter and a countdown timer. Each time an RFID tag is queried, the RFID tag will increment its counter after its timer expires. This is an atomic operation that cannot be disrupted. The intuition is for the reader to query each RFID tag one after the other to collect the responses to generate a proof before the timer expires.

Fig. 10.4 illustrates a grouping proof from Bolotnyy et al.[27]. The proof is to demonstrate that RFID Tag 1, Tag 2, and Tag 3 are present at the same time. To generate the proof, the reader will first query the first tag, Tag 1, and receive a_1 where $a_1 = \{id_1, c_1, h(s_1, c_1)\}$. At this point, the timer for Tag 1 has started. The reader will continue to send a_1 to Tag 2 and receive a_2 back (Step 2). The value of a_2 is $\{id_2, c_2, h(s_2, c_2, a_1)\}$. The reader will send a_2 to Tag 3 and get back a_3 (Step 3), which is $\{id_3, c_3, h(s_3, c_3, a_2)\}$. At this time, the reader has collected responses from all the tags, and will send a_3 back to the first tag, Tag 1. This has to be done before Tag 1's timer expires. If the reader is successful, the reader will obtain the message m, where $m = h(a_1, a_3, s_1)$. RFID Tag 1 will not respond if the timer expires. The reader will then submit the proof p to the backend server for verification, where $p = \{id_1, id_2, id_3, c_1, c_2, c_3, m\}$. Since the backend server knows the secrets for each of the ids, the backend server can determine whether c_1, c_2, c_3, and m are valid.

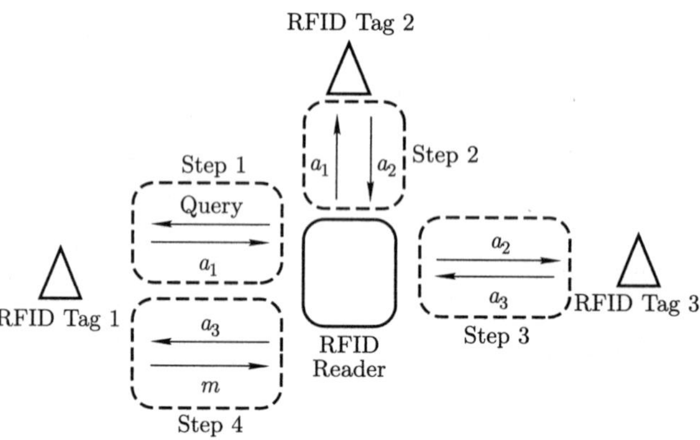

Fig. 10.4 Grouping proof for 3 RFID tags. After Step 4, the reader generates the proof, which is then transmitted to the backend server for verification. Steps 1 to 4 have to be completed before the RFID tag timer expires.

We can see that if the RFID reader does not complete the proof in time, the reader will be unable to return the correct m value to the backend server because computing m requires s_1, which is only known to the Tag 1 and the backend server. The reader also cannot reuse old values such as a_1, a_2, or a_3, since the counter value for each tag will increment each time, creating an incorrect m value. Grouping proofs are also known as "yoking-proofs", which was first proposed by Juels[28], which as limited to 2 tags. More recent work by Burmester et al.[29] and Tan et al.[21] improves on this concept.

10.7 Commercial RFID Security

In this section, we turn our focus to commercial RFID security solutions. Details regarding commercial RFID systems are often difficult to come by, since companies are reluctant to release information publically. Despite this, researchers have been successful in reverse engineering some RFID products. Recent work by Garcia, et al.[30], Kasper et al.[31], and Nohl, et al.[32] have demonstrated vulnerabilities in some commercial RFID systems.

Here, we consider the security mechanisms for RFID enabled passport (ePassport). Since passports have to be interoperable among various airports globally, documentation on the security mechanisms is available.

10.7.1 Background on RFID-enabled Passports

The standards for RFID-enabled passports are maintained by the International Civil Aviation Organization (ICAO), which maintain, among other things, the protocols needed to access the RFID tag embedded within passports. Since our focus is on RFID systems, we limit our discussion to the common interaction between the RFID reader and the tag. Details such as maintaining public key infrastructure (PKI) and RFID reader revocation are omitted. Interested readers can obtain more information from International Civil Aviation Organization documentation[33].

The RFID tag within the passport contains data relating to the passport holder, for instance, the height or photograph of the passport holder. Since the RFID tag has limited storage capacity, a hashed result of such information is stored in the RFID tag. The basic steps for verifying a passport is given in Fig. 10.5. These steps are performed, for instance, at the immigration counter at an airport.

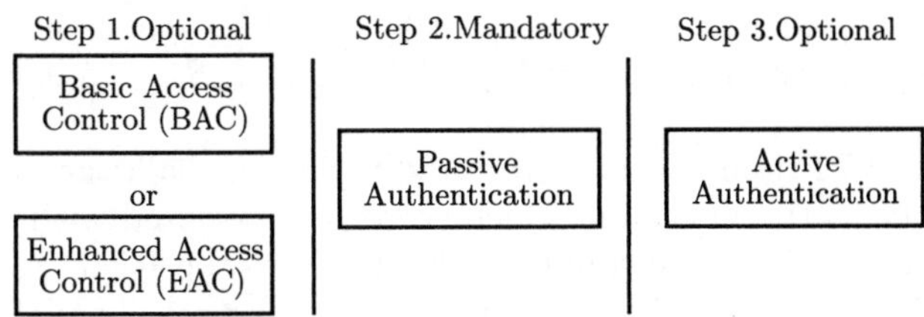

Fig. 10.5 Basic steps when RFID reader queries the RFID tag in the passport. At the end of Step 3, the RFID reader will determine whether the tag is valid or not.

The first step is supposed to regulate access to the data contained within the RFID tag. There are two types of access control, the basic access control (BAC) and the extended access control (EAC). According to Chothia et al.[34], most passports already implement BAC. We will focus our discussion on BAC in the next subsection.

The second step in the verification process is mandatory. The purpose of passive authentication is to verify that the data contained within the RFID tag is valid. When the passport is first issued, information about the passport holder is hashed and signed with a secret key that is associated with the country who issued the passport. In the passive authentication step, the hashed data and signatures are verified using the public key associated with the country.

The last step, active authentication, is needed because Step 2 only verifies that the data contained in the RFID tag is genuine. It does not indicate that the passport itself is legitimate. The reason is that an adversary could skim the data off the real passport, and stored it into the RFID tag of a fake passport. In Step 3, the RFID tag itself is authenticated. Performing step 3 requires the RFID tag to perform public key operations. In active authentication, the RFID reader will send a random number over to the RFID tag, which will then digitally sign the number and return the signature to the reader. Active authentication is also optional process.

A passport that only implements the mandatory passive authentication does not satisfy the three key RFID security requirements discussed earlier. From International Civil Aviation Organization document[33], the motivation for implementing BAC is to prevent skimming and eavesdropping. Even though preventing illicit tracking is not a stated goal of BAC, we will see in the later security analysis, BAC also protects against tracking.

10.7.2 Basic Access Control Protocol

An RFID tag that runs BAC has to be able to perform symmetric key operations. The tag will store two symmetric keys permanently, $K_{\rm enc}$ and $K_{\rm mac}$. These two keys are computed when the passport is first issued to the passport holder. The goal of BAC is to allow the RFID reader and the RFID tag to eventually derive a session key $KS_{\rm enc}$ and $KS_{\rm mac}$ to encrypt future transactions.

In the basic RFID protocol introduced earlier, a challenge in RFID protocols is to efficiently determine which secret is associated with a particular tag. A similar problem is encountered here, where the RFID reader has to determine which $K_{\rm enc}$ and $K_{\rm mac}$ belongs to the passport. The RFID enabled passport overcomes this problem by computing $K_{\rm enc}$ and $K_{\rm mac}$ using a function of

<Passport number, Passport holder's date of birth, and Passport's Expiry date>.

All passports must contain these three pieces of information. The reasoning is that these information can be easily obtained when a passport holder gives his passport to immigration personal for verification, upon which the RFID reader can obtain $K_{\rm enc}$ and $K_{\rm mac}$. Assuming that the RFID reader now posses $K_{\rm enc}$ and $K_{\rm mac}$, Fig. 10.6 shows the rest of the BAC protocol.

After the RFID reader issues a query, the tag will respond with a random number n_t. The reader will execute Step A, which consists of the following

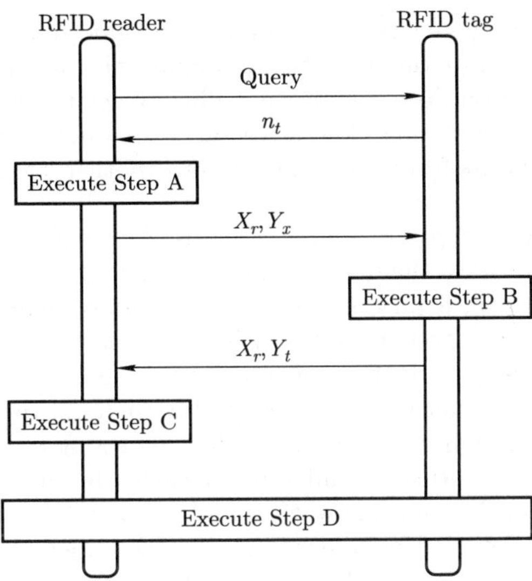

Fig. 10.6 Basic Access Control (BAC) for passport RFID tag.

substeps.

(1) Generate 2 random numbers, n_r and k_r.

(2) Compute $Z_r = n_r|n_t|k_r$

(3) Compute $X_r|Y_r$, where $X_r = E(Z_r, K_{\text{enc}})$ and $Y_r = h(X_r, K_{\text{mac}})$

After the tag receives $X_r|Y_r$, the tag will execute Step B. In this step, the tag will first verify Y_r using K_{mac}, and then decrypt X_r to obtain Z_r. The tag will check whether the n_t value in Z_r is the same as the value transmitted earlier. The tag will only continue executing if this n_t matches. The tag will finally compute $X_t|Y_t$ as follows.

(1) Generate random k_t.

(2) $Z_t = n_t|n_r|k_t$.

(3) Compute $X_t|Y_t$, where $X_t = E(Z_t, K_{\text{enc}})$ and $Y_t = h(X_t, K_{\text{mac}})$.

Upon receiving $X_t|Y_t$, the reader will execute Step C. Here, the reader will first verify that Y_t is valid using K_{mac}, and then decrypt X_t, and check whether the n_t value contained within Z_r is the same value it transmitted earlier.

Finally, in Step D, both the reader and tag will compute the session keys KS_{enc} and KS_{mac} using the value of K_r XOR K_t as seed. Subsequent communications between the reader and tag will be protected using KS_{enc} and KS_{mac}.

10.7.3 Security Analysis

A passport that only implements passive authentication does not meet the three security requirements. There is no access control mechanism, and thus any reader can query the tag and obtain the same information. Since the information returned by the passport is uniquely tied to the passport holder, passive authentication does not prevent illicit tracking. Finally, there is no protection against skimming. The adversary can simply query the tag, and then store the response from one passport onto another, to create a fraudulent passport.

The use of BAC can partly address these problems. Let us assume that the adversary does not have the information of K_{enc} and K_{mac}, i.e. the passport number, date of birth, expiry date of the passport is unknown to the adversary. The adversary querying the tag will be unable to get further than Step B (Fig. 10.6), since the adversary cannot return the X_r, Y_r values. Since all reader and tag communication is encrypted with K_{enc}, the adversary learns no information through eavesdropping. Illicit tracking requirement is satisfied because the tag returns a different n_t each time it is queried. Finally, since the adversary does not know K_{enc} and K_{mac}, a fake tag created by the adversary will be detected by a legitimate RFID reader.

Early work by Juels et al.[35] indicated the security pitfalls of implementing only passive authentication on passports, and advocated the use of BAC regardless of its limitations (early RFID enabled passports did not have BAC). More recent works have found practical security vulnerabilities for passports from specific countries[36]. BAC relies on the adversary being unaware of K_{enc} and K_{mac}, and work by Liu et al.[37] demonstrated such attacks. A practical tracking attack has been proposed by Chothia et al.[34] which can possibly be used to track passports based on the country of origin.

10.8 Conclusion

In this chapter, we studied the problem securing RFID networks and communications. The chapter focused on the weakest link, which is between the RFID reader and the RFID tag.

We described the characteristics of each of the components that make up an RFID network, and then categorized the security requirements for an RFID network. We first studied hardware based solutions to address these requirements. Then, we studied the more conventional protocol based approach towards RFID security. We studied protocols that can address the basic security requirements of preventing unauthorized access, illicit tracking, and skimming. We then turned our attention to security protocols that provide more advance security requirements of preventing mafia attack and providing grouping proofs. Finally, we concluded by studying a commercial RFID security protocol, the basic access control standard, used in passports.

We believe that RFID security will continue to be an important research area in the future as RFID is used in more critical applications. This chapter summarizes some of the key research in the security of RFID networks and communications, and we hope that our work can be used as a building block for future investigations into this problem.

References

[1] Gershenfeld N, Krikorian R, Cohen D (2004) The Internet of Things. Scientific American 291: 76 – 81.

[2] Koscher K, Juels A, Kohno T, Brajkovic V (2009) EPC RFID Tags in Security Applications: Passport Cards, Enhanced Drivers Licenses, and Beyond. In Proceedings of the 2009 ACM Conference on Computer and Communications Security (CCS'09), pp. 33 – 42.

[3] Kargoth G, Moskowitz P A (2005) Disabling RFID tags with visible confirmation: clipped tags are silenced. In Proceedings of the 2005 ACM workshop on Privacy in the electronic society (WPES'05), pp. 27 – 30.

[4] Moskowitz P A, Lauris A, Morris S (2007) A Privacy-Enhancing Radio Frequency Identification Tag: Implementation of the Clipped Tag. In Proceedings of the 2007 IEEE International Conference on Pervasive Computing and Communications Workshops (PerCom'07), pp. 348 – 351.

[5] Juels A, Rivest R, Szydlo M (2003) The Blocker Tag: Selective Blocking of RFID Tags for Consumer Privacy. In Proceedings of the 2003 ACM Conference on Computer and Communication Security (CCS'03), pp. 103 – 111.

[6] Floerkemeier C, Roland S, Marc L (2004) Scanning with a Purpose — Supporting the Fair Information Principles in RFID Protocols. International Symposium on Ubiquitous Computing Systems.

[7] Rieback M R, Crispo B, Tanenbaum A S (2005) RFID Guardian: A Battery-Powered Mobile Device for RFID Privacy Management. Australasian Conference on Information Security and Privacy.

[8] Juels A, Ravikanth P (2003) Squealing Euros Privacy Protection in RFID-Enabled Banknotes. Financial Cryptography and Data Security, 2742: 103 – 121.

[9] Cai S, Li Y, Li T, Deng R H, Yao H (2010) Achieving High Security and Efficiency in RFID-tagged Supply Chains. International Journal of Applied Cryptography 2(1): 3 – 12.

[10] Avoine G (2011) RFID Security & Privacy Lounge. http://www.avoine.net/rfid/. Accessed 14 June, 2011.

[11] Alomair B, Poovendran R (2010) Privacy versus Scalability in Radio Frequency Identification Systems. Computer Communication, Elsevier, 33(18): 2155 – 2163.

[12] Tsudik G (2006) YA-TRAP: Yet Another Trivial RFID Authentication Protocol. International Conference on Pervasive Computing and Communication.

[13] Chatmon C, Tri van L, Burmester M (2006) Secure Anonymous RFID Authentication Protocols. Florida State University Technical Report. http://www.cs.fsu.edu/~burmeste/TR-060112.pdf. Accessed 9 December, 2010.

[14] Tsudik G (2007) A Family of Dunces: Trivial RFID Identification and Authentication Protocols. In Proceedings of the 7th International Conference on Privacy Enhancing Technologies (PET'07), pp. 45 – 61.

[15] Hopper N J, Blum M (2001) "Secure Human Identification Protocols." International Conference on Theory and Application of Cryptology and Information Security (ASIACRYPT).

[16] Juels A, Weis S (2005) Authenticating Pervasive Devices with Human Protocols. Advances in Cryptology (CRYPTO).

[17] Piramuthu S (2006) HB and Related Lightweight Authentication Protocols for Secure RFID Tag/Reader Authentication. Collaborative Electronic Commerce Technology and Research.

[18] Holcom D, Burleson W, Fu K (2007). Initial SRAM state as a Fingerprint and Source of True Random Numbers for RFID Tags. Conference on RFID Security.

[19] Melia-Seguil J, Garcia-Alfaro J, Herrera-Joancomarti J (2010) Analysis and improvement of a pseudorandom number generator for EPC Gen2 tags. International Workshop on Lightweight Cryptography for Resource-Constrained Devices.

[20] Peris-Lopez P, Millan E S, van der Lubbe JCA, Entrena L A (2010) Cryptographically secure pseudo-random bit generator for RFID tags. International Conference for Internet Technology and Secured Transactions.

[21] Tan C C, Sheng B, Li Q (2010) Efficient techniques for monitoring missing RFID tags. IEEE Transactions on Wireless Communication 9(6): 1882 – 1889.

[22] Tan C C, Sheng B, Li Q (2008) Secure and Serverless RFID Authentication and Search Protocols. IEEE Transactions on Wireless Communication 7(4): 1400 – 1407.

[23] Hancke G P, Kuhn M (2005) An RFID Distance Bounding Protocol. Conference on Security and Privacy for Emerging Areas in Communication Networks.

[24] Avoine G, Muhammed A B, Suleyman K, Cédric L, Benjamin M (2010) A Framework for Analyzing RFID Distance Bounding Protocols. J. Comput. Secur. 19(2): 289 – 317.

[25] Kim C H, Avoine G (2009) RFID Distance Bounding Protocol with Mixed Challenges to Prevent Relay Attacks. International Conference on Cryptology And Network Security.

[26] Capkun S, El Defrawy K, Tsudik G (2010) GDB: Group Distance Bounding Protocols. arXiv.org, http://arxiv.org/abs/1011.5295. Accessed 15 January, 2011.

[27] Bolotnyy L, Gabriel R (2009) Generalized "Yoking-Proofs" and Inter-tag Communication. In Development and Implementation of RFID Technology. I-Tech Education and Publishing.

[28] Juels A (2004) "Yoking-Proofs" for RFID Tags. International Workshop on Pervasive Computing and Communication Security.

[29] Burmester M, de Medeiros B, Motta R (2008) Provably Secure Grouping-Proofs for RFID Tags. Smart Card Research and Advanced Applications (CARDIS).

[30] Garcia F, Koning G G, Muijrers R, Rossum P, Verdult R, Schreur R W, Jacobs B (2008) Dismantling MIFARE Classic. European Symposium on Research in Computer Security (ESORICS).

[31] Kasper T, Silbermann M, Paar C (2010) All You Can Eat or Breaking a Real-World Contactless Payment System. Financial Cryptography and Data Security. 6052: 343 – 350.

[32] Nohl K, Evans D, Starbug, Plotz H (2008) Reverse-Engineering a Cryptographic RFID Tag. USENIX Security Symposium.

[33] International Civil Aviation Organization (2009). Doc 9303 Part 1 Vol 1 and 2. http://www2.icao.int/en/MRTD/Pages/Downloads.aspx. Accessed 15 January 2011. Accessed 15 January, 2011.

[34] Chothia T, Smirnov V (2010) A Traceability Attack against e-Passports. Financial Cryptography and Data Security, 6052: 20 – 34.

[35] Juels A, Molnar D, Wagner D (2005) Security and Privacy Issues in E-passports. Conference on Security and Privacy for Emerging Areas in Communication Networks.

[36] Avoine G, Kassem K, Jean-Jacques Q (2008) ePassport: Securing International Contacts with Contactless Chips. Financial Cryptography and Data Security, 5143: 141 – 155.

[37] Liu Y, Kasper T, Lemke-Rust K, Paar C (2007) E-Passport: Cracking Basic Access Control Keys. OTM confederated international conference on the move to meaningful Internet systems: CoopIS, DOA, ODBASE, GADA, and IS - Volume Part II.

Index

D

E

F

G

H

I

J

K

L

M

N

O

P

Q

R

S

T

U

V

W

Z